STO

OUR
CHANGING
UNIVERSE

John Gribbin

OUR CHANGING UNIVERSE

THE NEW ASTRONOMY

A Sunrise Book
E. P. DUTTON & CO., INC.
New York

For Minna
To show her what I've been doing
the past few years

Half-title page: The rising earth seen from lunar
orbit with the lunar landing module in the foreground.
Title page: M51, the Whirlpool Galaxy.

Acknowledgements

The publishers are grateful to the following for supplying
illustrations:

Bell Labs: 53, 122–3, 123 inset; Copyright, November 1974, Bell
Telephone Laboratories Incorporated. Reprinted by kind permission,
Editor, Bell Labs Record: 52; British Insulated Calender's Ltd: 18;
Catalina Observatory: 103 inset; European Space Agency: 111t, 111b;
R F Carswell: 26; Ray Davis Jnr: 73; Durham University: 56, 57;
Glasgow University: 116–17; R S Hey: 15; Gary Ladd: 106–107;
Leiden Observatory: 13; Lick Observatory: 2, 36–7, 38, 67, 70–71,
74t & b, 75t & b, 78–9, 96, 102, 103t, 110, 128, 130, 131, 134;
Manchester University: 8, 9; Max-Planck-Institut: 12; Mullard Radio
Astronomy Laboratory, Cavendish Laboratory, Cambridge: 19;
NASA: 1, 137t, 137b, 138b, 139t, 141, 143, 145l, 145r, 103b;
NCAR Photo: 60–61; Novosti Press Agency: 10–11; Pennsylvania
State University: 68; Radcliffe Observatory: 47; Royal Astronomical
Society: 23, 138t; Royal Greenwich Observatory: 76, inset 76;
Science Research Council: 45, 50–51, 100, 104–105, 112,
126, 133, 149tL, 149bL, 149tr, 149br, 151, 152; *Sky and Telescope*
(Dennis Milon): 66; Stearn's, Cambridge: 86; US Geological Survey:
139m, 139b

First published in the US 1976 by E. P. Dutton & Co., Inc.

Copyright © 1976 by John Gribbin
All rights reserved. Printed in Great Britain by
Thomson Litho Ltd., East Kilbride, Scotland
First Edition
10 9 8 7 6 5 4 3 2 1

ISBN: 0–87690–216–6
Library of Congress Catalog Number: 76–41–29

Contents

Introduction

In the past ten years or so, a series of dramatic developments has changed the face of astronomy. Radio astronomy had, of course, already become well established by the early 1960s, providing a new way to study the known universe. But with the series of discoveries beginning with the identification of the first quasar, 3C 273, the new astronomy began. With that identification radio astronomy was no longer telling us new things about the universe we already knew—but was beginning to reveal the existence of previously unsuspected objects within that universe. The discovery of the so-called 'cosmic background radiation' provided a new insight into the very processes which formed the universe; while by 1968 pulsars had been found, demonstrating that our ideas had to be revised even when applied to objects in our own cosmic backyard, our own galaxy.

While these developments were occurring in radio astronomy, a series of rocket-borne detectors, carried above the atmosphere, revealed the presence of celestial sources of X-rays. In the early 1970s, the first X-ray astronomy satellites showed that the number of these X-ray stars runs into hundreds, and again new ideas had to be developed to accommodate these discoveries. Exotic objects such as black holes were called in to explain the production of energy in the X-ray stars.

This dramatic change in astronomers' view of the universe, in such a short space of time, is the result of opening up new parts of the electromagnetic spectrum to study. Radio waves, which penetrate the atmosphere, provided a new 'window' on the universe. With rockets and satellites, even those wavelengths, like X-rays, which are stopped by the atmosphere could be examined. Every part of the electromagnetic spectrum can now be observed, and that is why there are unlikely to be the same kind of dramatic new discoveries —quasars, pulsars, and X-ray stars—in the future. Gravitational radiation might perhaps provide another window through which we can observe the changing universe, and there have been claims that such radiation has already been detected. There will be many exciting developments too, as these techniques are improved. But now, when man has just opened wide the observational door and has taken his first astonished look at the unsuspected wonders revealed, seems the ideal time to look back on the most exciting ten years in the history of astronomy. Our universe itself is changing and evolving; equally fascinating is the way in which our understanding of the universe is changing in the light of the new astronomy.

1 Radio Astronomy

The story of radio astronomy is the oldest and most familiar of all the developments which will be mentioned in this book. Several excellent volumes have been written about this part of the new astronomy alone; the brief account given here is not intended to replace them, but rather to indicate how radio astronomy ties in with the other aspects of the new astronomy.

Like so many other important advances in science, radio astronomy came into being chiefly through inspired interpretation and adaptation of a discovery which, if not quite an accident, was certainly unexpected. It was as long ago as 1930 that Karl Jansky, a physicist working for the Bell Telephone Laboratories in the USA, was set the task of investigating the origin of atmospheric static, or 'radio noise', which was proving troublesome to the new radio communications industry. Jansky found that one kind of static could be explained by the influence of local thunderstorms, and that a weaker but steadier background was probably due to the average effect of many distant storms, with their lightning flashes and electrical activity. But there was also a third kind of noise.

That noise might have aroused little interest in someone else, and at first Jansky thought it was caused by man-made interference. But his interest was stirred when he noticed that this third source of radio noise was directional—it moved around the sky as the Earth rotated. It seems obvious now, with hindsight, to say that the source of the noise must have been outside the Earth, but Jansky's immediate recognition of the importance of his discovery ranks with Newton's realisation that something important could be learned by watching an apple fall. Further study showed that Jansky's equipment was picking up a background hiss of radio noise from the centre of our own galaxy. Radio astronomy had been born. In fact it was to show little development for ten years, and only became 'big science' after the Second World War had hastened the development of radio and radar techniques and equipment which could be turned to astronomical applications.

To most people, mention of radio telescopes immediately conjures up a picture of the Mark I instrument at Jodrell Bank, and the story of Jodrell Bank is an appropriate way to pick up the story of how radio astronomy became big science. The idea for this huge steerable aerial, 250 feet in diameter, was conceived by Bernard Lovell in 1947; he proposed the building of a great radio telescope, which would be to radio astronomy what the 200-inch telescope at Mount Palomar in the USA was to optical astronomy. It was largely

Above: The Mark I radio telescope at Jodrell Bank

thanks to Lovell's enthusiasm, and his ability to raise funds and instil similar enthusiasm in others, that the project ever became more than a dream. Probably even Lovell, however, would have been daunted if he had realised at the outset what he was getting into. It was ten years before the telescope became the operational reality of Lovell's hopes, and the cost, at getting on for a million pounds, was ten times the original estimates. Indeed, the embarrassing escalation in cost cast serious doubts on the future of the telescope—until, by another remarkable stroke of serendipity, the first Sputnik was launched in 1957. By tracking this and subsequent satellites Jodrell Bank brought publicity and prestige to the radio astronomy community.

But, despite the publicity and the way Jodrell Bank has captured the popular imagination, large steerable instruments are not the only ones used by radio astronomers. In Britain, the other centre of radio astronomy is at Cambridge, and there developments have centred around the interferometry techniques, by which two or more smaller aerials (ranging up to whole fields covered in aerials, or a series of larger aerials spaced down a railway track) can be used to mimic some of the properties of giant aerials miles across.

With the simplest kind of interferometer, the radio signal received by two connected aerials is interpreted electronically as if the two aerials were part of one large array covering the whole area between the aerials. If you had a large square field, marked out in a grid like a chessboard, you could build up an average picture of a radio source by moving your two aerials about from square to square on the grid and making a series of observations; this is the basis of the idea of aperture synthesis, in which a response like that of a large aperture telescope is 'synthesised' from observations with small telescopes. The Cambridge group, under Martin Ryle (now the Astronomer Royal), have developed this to a fine art, and life is made a lot easier because most of the labour of moving the two aerials around can in fact be avoided. The trick is simply to allow the Earth's rotation to move the aerials for you, since what matters is not the position of the aerials on the Earth, but their relative positions in space with respect to the radio source being studied.

Interferometry and synthesis techniques have become a vital tool of radio astronomers around the world. At Jodrell Bank, in Australia, in the USA and elsewhere the tool has now been developed to the point where it is possible to link up instruments on different continents, thus providing baselines as wide as the diameter of the Earth itself. Such techniques give very detailed information about the structure of radio sources. More commonly, since such worldwide cooperation is difficult to set up, links are established between different radio astronomy establishments in the same country. In Britain, for example, the long baseline between Jodrell Bank and the Royal Radar Establishment aerials at Malvern is one which has been used with a great deal of success.

Below: The Mark II radio telescope at Jodrell Bank, used in conjunction with the Mark I and other instruments for interferometry studies

But the development of ingenious systems for radio observation has by no means been a British preserve. In Australia, B. Y. Mills developed an aerial system in the form of two long arrays forming a cross on the ground; the system, logically enough, is called a Mills Cross, and can be used to simulate a very narrow beam, a pinpoint radio 'eye' reaching upward from the intersection of the cross. The Australians also have large steerable aerials, and in the Soviet Union and the USA there are aerials of all kinds in use at radio astronomy observatories. In radio astronomy terms the Netherlands is among the leaders: at Westerbork, a system of twelve antennas is combined in a mile-long array which is used to simulate the resolving ability of a radio telescope one mile in diameter.

In the summer of 1974, astronomers working with the Westerbork telescope reported the discovery of two giant radio sources, much larger in area than any previously identified. This opened the way for a whole new field of radio astronomy studies. The two sources are code-named 3C236 and DA240. Before the Westerbork study, radio astronomers had thought that they were each complexes of a

Below: Part of a large 'T' array in the Soviet Union. This instrument works on the same principle as a Mills' Cross

great many overlapping but unrelated objects. But the great resolving power of the Dutch telescope showed that each is a single, huge, sprawling radio source.

Astronomers measure the *apparent* size of a celestial object to an observer on earth in terms of the angle or degrees of arc of the sky covered. One degree is divided into 60 minutes (arc min), and one minute into 60 seconds (arc sec). The *actual* size or distance of an object can be measured in various units, as for example light years, or the parsec—a unit of distance equal to 206,265 times the distance of the Earth from the Sun, or about 3·26 light years.

Now, it seems clear that each of these sources covers more than 30 minutes of arc on the sky—a bigger angle than that covered by the apparent diameter of the Moon. Allowing for the immense distance of these objects, that means that the region of radio emission associated with 3C236 is 5·7 Megaparsecs (5·7 million parsecs) across, and the size of the DA240 radio source is at least 2 Megaparsecs. Our own galaxy is only about 30,000 parsecs across, so 3C236 is more than 150 times as extensive as our galaxy.

It remains to be seen whether these two giant radio sources are freaks, or whether they are typical of a previously unsuspected kind of radio source. Either way, the evidence they provide of interactions between galaxies, and of the nature of the material of intergalactic space, is proving of great value to astronomers interested in the intergalactic medium. For, of course, a radio source 150 times as big as our galaxy is not even as 'solid' as our galaxy. Most of the extent of the radio source is made up of a good approximation to what we would regard as empty space. The radio noise is coming from a scattering of electrons, probably ejected from a more compact central galaxy, which are interacting with the tenuous gas and magnetic fields of intergalactic space.

Even with such large sources, however, it is the detailed structure which is likely to provide most information about the physics of the processes by which the radio noise is produced. And this, of course, is where the very long baseline interferometry techniques—even intercontinental radio astronomy—become so important.

The resolution of a telescope—the amount of detail it can see—depends on the ratio of the aperture of the telescope to the wavelength of the signal being received. Because radio waves are so much longer than light waves, a 'telescope' as big as the Earth is needed to resolve radio sources with as much detail as practicable optical telescopes can give of stars and galaxies. And, surprising as it may seem, this can be achieved, in effect, by adding together the signals

Below: The 100 metre diameter fully steerable radio telescope of the Max Planck Institute for Radio Astronomy. This is the largest instrument of its kind at present operating anywhere in the world

Right: In this picture contour lines of radio emission have been superimposed on a photograph of the Spiral Galaxy M51. Regions of strong radio emission, like the bright stars, follow the spiral arms

from widely spaced radio telescopes.

In a radio interferometer, the wave from a distant object reaches two separated radio telescopes and is analysed by comparing the two signals electrically. In principle, the same kind of technique would work for optical telescopes as well; but the comparison depends on measurements accurate to a fraction of a wavelength, so in this case the longer wavelength of radio signals is an advantage! The biggest optical stellar interferometers are only a score of feet or so across.

The first radio interferometer telescopes were linked by cables, which provided obvious difficulties when it was wished to extend the separation of the two elements beyond a few kilometres. Microwave radio relay links, used in pioneering studies by British and Australian radio astronomers, pushed the separation of the two antennas in a radio interferometer out to 100 km for the first time, and this gave a resolution of better than 1 arc sec (for comparison, the Moon covers about half a degree on the sky, that is 30 arc min, or 1800 arc sec). But the ultimate step—at least, ultimate as long as radio astronomers are restricted to work on Earth—depended on cutting out the direct link between the telescopes altogether.

Today, radio telescopes on opposite sides of the Earth are used to record the signals from a distant source, and the two tape-recordings are then compared. The trick, of course is that the timing of the recordings must be very precise, so that astronomers can be sure that they are adding together the effects of radio waves which really did reach the Earth simultaneously. The actual synchronisation must be to within one millionth of a second, but in practice the original synchronisation need only be one tenth as good.

Even that is quite an achievement, involving the use of the latest atomic clocks, which measure time by the oscillations of caesium or rubidium resonators. Two identical oscillators have to be set up together, then taken to the two radio telescopes where the recordings are to be made. Then, the two resulting tapes are taken to the same laboratory for analysis. With the 'approximate' synchronisation provided by the atomic clocks ('only' accurate to one hundred-thousandth of a second or so), it is fairly straightforward, using modern computers, to align the two recordings precisely by comparing the details of the electromagnetic signals. Then the signals can at last be analysed.

Some idea of the power of this technique can be grasped from this analogy given by Professor F. Graham Smith (now Director of the Royal Greenwich Observatory) in his book *Radio Astronomy*. Several quasars are now known to have angular diameters of less than one thousandth of a second of arc, which is the size a tennis ball would appear if viewed at a distance equivalent to the diameter the Earth. The best resolution obtained by radio interferometry so far is five times more detailed still, at 200 millionths of a second of arc—a fineness of resolution which makes a dramatic contrast to the huge area, more than 30 *minutes* of arc, spanned by both 3C236 and DA240.

And there is still room for improvement, if ways could be found of correlating the signals of many different radio telescopes around the world, instead of just using two at a time. The mind-boggling concept of aperture synthesis applied on a literally global scale could yet become a reality. But individual telescopes (or more modest aperture synthesis arrays) are still the principal tool of radio astronomy, and intercontinental radio astronomy is still very much the exception rather than the rule.

Of all the instruments now operating, perhaps one more should be mentioned specifically—the radio 'telescope' at Arecibo, where a natural depression in Puerto Rico has been bulldozed into a more-or-less smooth surface and lined with reflecting metal. The radio mirror resulting is one kilometre across, the largest of its kind. Although this telescope is not, of course, steerable, and can only look at one band of the sky as the Earth sweeps it round, there is plenty to look at in that band of the sky, and the Arecibo observatory plays its part in revealing the nature of the changing universe.

These are just some of the instruments and techniques used; the examples are not intended to be comprehensive. But what exactly is it that these instruments, and others like them, tell us about the universe? How has our understanding of what goes on in space changed as radio astronomy has developed? The best way to get an insight into the answers to these questions is to look at one particular example, to see how radio astronomy has changed our understanding of one particular celestial object.

2 The Story of Cygnus A

Cygnus A was one of the first sources of radio noise to be discovered to lie far outside our own galaxy.

The radio source itself was first detected in 1946 by a group led by J. S. Hey and working in Richmond Park; the aerial they used looks almost ridiculous compared with the giant dish at Jodrell Bank, or modern interferometers, and was scarcely more complicated than a modern multi-element (yagi) array of the kind used for domestic television reception. This equipment could only locate the position of this powerful radio source to within about 2 degrees of arc on the sky—enough to show that it lay in the direction of the constellation Cygnus (hence its name), but not enough to permit identification with any definite optical object.

Apart from its great strength, the important thing about the radio emission from Cygnus A was that it fluctuated slightly, in a way which looked superficially very much like the variations of the radio noise from the Sun. Naturally enough, astronomers decided that Cygnus A must be a powerful radio star. This discovery provided a

Below: Surplus radar equipment used by J S Hey in the first studies of Cygnus A from Richmond Park in 1946

great boost for radio astronomers, and the early interferometers were used to measure the size and position of Cygnus A with more accuracy. But the big surprise came in 1951, when the true nature of Cygnus A was revealed. It was identified, not with a star in our own galaxy, but with a galaxy 550 million light years away from us.

The identification, which provided another breakthrough in radio astronomy, was made by F. Graham Smith, using the most sophisticated interferometer methods then possible. In fact, this work was part of Smith's research thesis, for which he was awarded the first Ph.D. in radio astronomy from the University of Cambridge; Smith is now Director of the Royal Greenwich Observatory at Herstmonceux. The accurate interferometer study showed that the radio source Cygnus A probably coincided with an unnamed galaxy. The 200-inch telescope at Mount Palomar was then used to photograph the galaxy, and it turned out to be a very unusual object indeed. The evidence was overwhelming—this peculiar galaxy, clearly involved in some cataclysmic process, must indeed be the source of the strong radio signals detected for the first time five years before.

The discovery plunged radio astronomers into a complete revision of their ideas about the radio universe. Cygnus A was the second strongest source of radio noise known, and yet it could be identified with a very remote galaxy. The distance of other galaxies can be found quite easily because of the expansion of the universe; since the galaxies are moving away from one another, the lines in their optical spectra are redshifted by the Doppler effect and the magnitude of the redshift reveals their distances. We shall learn more of the redshift and the expansion of the universe when we come to quasars and cosmology. Here, we need only try to picture the consternation caused to the students of the new science of radio astronomy when they found out just how far away Cygnus A really is, and realised that to be as 'bright' as it is at radio frequencies, it must be a million times brighter in the radio part of the spectrum than an ordinary galaxy like our own.

What could be producing this intense radiation? The photographs obtained with the 200-inch telescope had shown that the Cygnus A galaxy—which by now had been given the name already attached to the radio source—had an unusual double structure. This at first misled the theoreticians, who suggested that the radio noise was produced by a collision between two galaxies. But as more radio galaxies were discovered it became clear that this explanation just would not do. A collision between galaxies is a very rare event indeed; but dozens of radio galaxies were soon identified. And the ever improving radio observations themselves paved the way for new and better theoretical 'models' to explain the nature of Cygnus A.

The problem of studying the detail of such remote objects is that they occupy such a very small angle on the sky, about 1 minute of arc in the case of Cygnus A. It seems remarkable enough that the source can indeed be located and identified with a particular galaxy, let alone mapped and studied in detail. But that is where interferometry comes into its own. By 1953, work at Jodrell Bank had shown that, like the optical galaxy, the radio source Cygnus A was actually a double source—but the two radio components were further apart than the centres of optical brightness. Because the two strong radio components are equally spaced either side of the optical galaxy, the earlier measurements had given the impression of a single radio source centred on the optical galaxy; whereas actually the radio sources are each some 44 arc sec away from the bright galaxy, so that the whole system really spans rather more than 1 minute of arc— something nearer to 1·5 arc min, in fact.

Now it was the turn of the theoreticians to think again. Each advance of radio astronomy technique posed new puzzles, and as fast as theories came along the radio astronomers developed still better techniques, producing observations which sometimes helped the theories along, and sometimes showed them up as completely ill-founded. In the late 1950s the likes of Geoffrey Burbidge, Fred Hoyle, and the American Willy Fowler developed the idea that radio galaxies are formed by vast explosions at the centres, or nuclei, of galaxies. Particles are shot out of the galaxies, and interact with magnetic fields to produce the radio emission we see as a double source embracing an optical galaxy. As more and more double radio sources like Cygnus A were discovered, and as the optical photographs were produced showing clear evidence of violent processes going on at the centre of some galaxies, this idea became firmly established, and now forms a cornerstone of astronomy. But it is only a broad outline, and many questions remain to be answered. How do these explosions occur? Can they occur in any galaxy or only a select few? And so on.

In a sense, the story of Cygnus A had by now come full circle. The idea that the central galaxy was the source of the radio noise had been replaced by the discovery of the double lobe of radio emission from either side of the galaxy; but now the theoreticians were saying that the radio outbursts had their origin in a central explosion in the galaxy itself. Surely such an explosion should leave behind some trace, perhaps in the form of a weaker 'radio-simmering' at the site of the ancient explosion? That certainly looked plausible. But if it was difficult to detect the presence of two strong radio sources with their centres of activity about 1·5 arc min apart, it was a much harder task to find a weaker radio source lying halfway between them. Indeed, it was only in 1973 that the radio noise from the centre of galaxy Cygnus A itself was measured for the first time.

The Jodrell Bank and Cambridge University groups each made observations of this central radio component in Cygnus A early in 1973; both used interferometry techniques, although there were considerable differences between their two sets of equipment, differences which show how far radio astronomy has come since Cygnus A was first noticed. The Jodrell interferometer used on this occasion consisted of two aerials, the Mark II and Mark III radio telescopes, which are 24 km apart. Both of them have elliptical parabaloid sur-

faces, the Mark II a solid one and the Mark III a wire mesh surface. The Mark II is alongside the famous Mark I telescope, at Jodrell Bank proper, while the Mark III, 24 km away, can be controlled remotely from Jodrell Bank.

The amount of detail which can be picked out from a distant source by an interferometer depends, as we have seen, both on the separation of the aerials (the baseline) and on the wavelength at which they are operating. The important factor is the number of wavelengths which would make up a length corresponding to the interferometer baseline. At 408 MHz, for example, the Mark II/ Mark III separation would correspond to 32,000 wavelengths, and the system could resolve details at least down to 7 arc sec across; the detailed mapping of Cygnus A, however, was at 1660 MHz, where the separation corresponds to 130,000 wavelengths, and details down to 1·5 arc sec across can be picked out. This system was actually completed in 1968; the 1973 study of Cygnus A was carried out by Dr R. J. Peckham who found that there is indeed a definite faint source of radio noise centred on the galaxy in Cygnus A, but that this source is only one thousandth as strong as each of the strong sources which together produce the radio emission that first drew attention to the system.

This is, perhaps, a good place to mention some other interferometer systems which have one 'foot' at Jodrell Bank, as it were. One commonly used link is with the Royal Radar Establishment at Malvern, 127 km away, where there is an 85-foot dish aerial; at 1660 MHz this baseline corresponds to 690,000 wavelengths, pro-

Left: Aerial view showing five of the eight dishes which make up the '5km' radio telescope near Cambridge. Below: Radio map of Cygnus A obtained with the '5km' telescope showing the 2 main lobes and the weak central source. The shaded oval shows the beamwidth of the telescope at the operating frequency, 5GHz

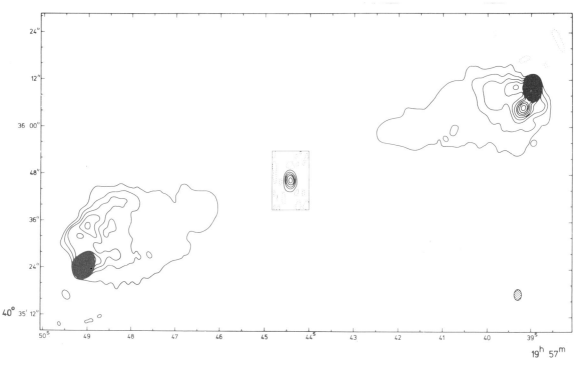

19

ducing a resolution of $\frac{1}{3}$ arc sec, and the system has been used at 5 GHz (5000 MHz) when the baseline corresponds to two million wavelengths and resolution down to 0·05 arc sec is possible.

There is, however, a snag with interferometers which have one fixed baseline—they cannot produce a complete 'map' of the radio contours of a source, but only a series of strips which show the structure along certain directions. The latest British radio telescope system overcomes this problem, although it suffers from another handicap, having a relatively short maximum baseline of about 5 km. The most information is obtained, of course, by studying one source with both kinds of system (and indeed with completely different systems, when they are available). So the recent Cambridge work provides an ideal complement to the Jodrell Bank study.

The new Cambridge telescope is called the 'Five Kilometre', because of its size. It is built along an old railway track and consists of four fixed aerials and four movable aerials spaced along 4·6 km of track (not quite 5 km, in fact!), which is aligned exactly East-West.

With these aerials, 16 interferometer spacings are available at any one time, and by moving the four mobile aerials along the track another 16 spacings come into play. In the 1973 study of Cygnus A, four sets of observations, each for 12 hours, were made giving a total of 64 spacings. This allows mapping of a region nearly 3 arc min across, comfortably covering the whole Cygnus A system. Like the Jodrell Bank team, the Cambridge group found the weak radio emission from the galaxy itself. The mapping revealed features down to a couple of seconds of arc across, because although the baselines were all shorter than the Jodrell 24 km, the operating frequency was much higher (at 5 GHz), so that about the same number of wavelengths at this frequency would be required to bridge the baseline as would be required at 1660 MHz to bridge the Mark II/Mark III baseline.

That is where things stand at present. More detail than ever before is available about Cygnus A (and about other radio sources) and while the explosion theory still looks good there is now plenty more detail for it to explain. Of course, radio techniques will continue to improve, and it does not require too much imagination to foresee the development of larger variable-spacing interferometers which combine in one system the virtues of the Jodrell Bank and Cambridge interferometers.

But, by and large, radio galaxies are beginning to be understood; they are familiar objects, dating back in man's knowledge to the dawn of radio astronomy, thirty years ago. Familiarity has perhaps bred contempt, and radio galaxies may have surprises yet to spring on the complacent. But in recent years the spotlight has turned to other kinds of object. One, the quasar, now seems to be related in some way to galaxies; the other, the pulsar, provided a surprise for radio astronomers in our own astronomical backyard, the Milky Way Galaxy itself; and both serve as a caution against complacency. So perhaps we should leave the story of Cygnus A only with the thought in mind that it may yet turn out to be far from over. In any case, the story so far has highlighted the development of radio astronomy in three decades—from a science capable only of saying 'there is something peculiar out past the constellation Cygnus', to one capable of providing measurements of the structure of one small part of that something peculiar, at a range of 550 million light years.

3 Quasars

Throughout the 1950s and into the 1960s radio astronomy was growing and flexing its new muscles. But exciting though the discovery of radio galaxies was, radio astronomy was, by the early 1960s, still 'only' telling astronomers more about objects they already knew. This situation was dramatically changed on March 16, 1963, with the announcement that the radio source 3C 273 had been identified with a starlike object which showed a redshift of 0·158. With that discovery, radio astronomy truly came of age, for that object was, as it turned out, the first identified example of a completely unknown category of astronomical object—the quasars. Today it is known that about 25% of all known celestial radio sources are quasars, and the investigation of these objects has caused a revolution in astronomy and cosmology.

To see the reason for this, we have to consider the two characteristic properties of quasars, which were clearly apparent in 3C 273. They are starlike (hence their name, from '*quasi-stellar* source, or object'), yet they show a large redshift in their spectra. This combination of properties seemed almost heretical to the ordered arrangement of the universe as understood in 1963. The cornerstone of cosmology as an observational science was, and indeed still is, Hubble's law concerning the recession of the distant galaxies. This law simply states that all the galaxy clusters outside our own seem to be receding from us, and that their velocity of recession is proportional to their distance from us. That does not mean that our own galaxy is at the centre of the universe, of course; the situation is similar to the view you might get from one plum in an expanding plum-pudding— every other plum would be receding from you, whichever plum you happened to be sitting on.

Hubble formulated his law some fifty years ago, to explain the fact that the light from distant galaxies is shifted towards the red end of the spectrum. This effect is measurable since the position of characteristic lines in the spectrum can be measured, and the simplest explanation is that it is a 'Doppler effect', caused by movement of the source of the light away from us. Much the same effect produces a deepening in the note of an ambulance or police siren as we are overtaken by the vehicle making the noise.

Now, Hubble's law worked very well in explaining what was happening to galaxies in the expanding universe, and we shall see just how it is used by cosmologists later on (Chapter 14). But the discovery of quasars threw the whole cosmological scene into turmoil.

First of all, why was it only in 1963 that such an important discovery was made? To a non-astronomer it sometimes comes as a

Right: The quasar 3C273 showing the bright jet (outlined in white on the photograph) which is also a strong radio source

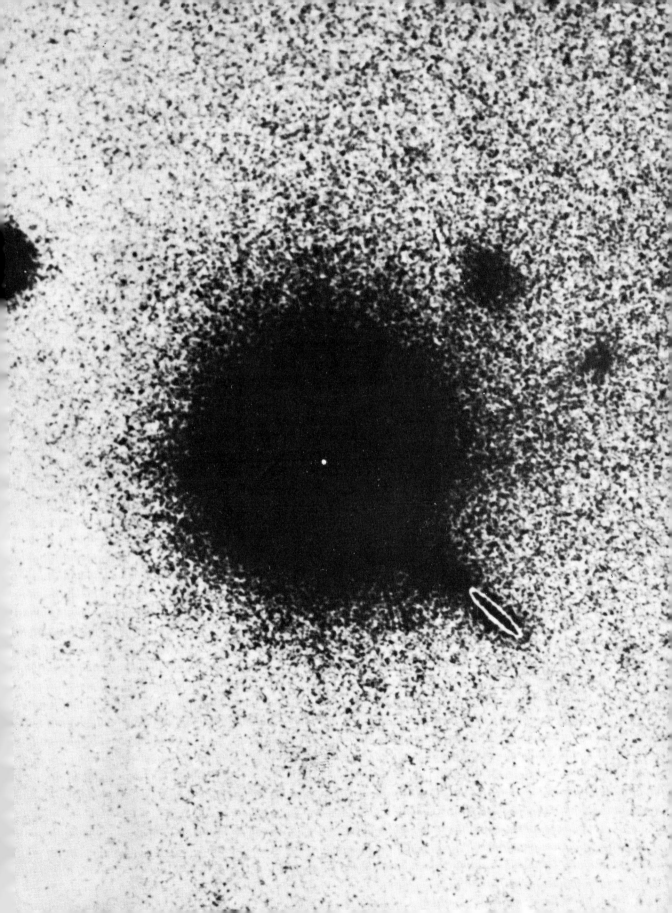

surprise to learn how difficult it is to pin down the exact position of celestial objects. We have already seen how vague the first indications of the positions of radio sources were; and even though astronomers knew roughly where to look for an optical counterpart to 3C 273, they were, by the early 1960s, strongly influenced by the discovery of radio galaxies. The last thing that they would suspect of being associated with a powerful source of radio noise would be a faint blue 'star' visible on the photographic plates of the region of the sky where the radio noise was coming from.

So the identification had to wait for very precise radio measurements—radio measurements so accurate, in fact, that only one object on the photographic plates could possibly be associated with the radio source. This was achieved using a technique known as lunar occultation.

A lunar occultation is simply the 'eclipse' of any celestial object as the Moon moves in front of it. Now, the Moon doesn't actually occult very many celestial objects; it is fairly small and moves round one band of the sky only, as viewed from the Earth. But when it does occult something interesting, astronomers can make good use of the event.

The point is that the exact position of the Moon on the sky at any instant is known very precisely, because the Moon follows a very regular orbit. When the Moon moves in front of, say, a radio source, the signal is cut off; and at the precise moment that the signal disappears the astronomers know precisely where the 'leading edge' of the Moon was. That tells them that the interesting object lies just on the semicircle corresponding to the leading edge of the Moon at that time. When the Moon has moved past the radio source, the occultation ends and the radio signals are picked up again. Now, the astronomers draw a second arc corresponding to the position of the 'trailing edge' of the Moon at the critical instant. The two arcs will cross at two points on the sky, but usually only one of these points is near to the approximate position of the source already known. And that means that the source has been pinned down to an accuracy of better than 1 arc sec.

That was the technique which produced the unambiguous identification of 3C 273. But how were the observations to be explained? If Hubble's law was applied to the redshift of 3C 273 then it seemed that the object must be well outside our galaxy—as far away as many other galaxies. But the quasar is so small that if it is visible at such a distance an enormous amount of energy must be produced from an object much more compact than any galaxy. The problem became worse when other quasars were found, with larger and larger redshifts. Soon, 3C 273 seemed almost like a close neighbour and quasars with redshifts up to about 2 (more than ten times the redshift of 3C 273) were known. Could these really be objects much smaller than galaxies but emitting as much power as all the stars of a large galaxy put together?

Some astronomers doubted this. But the alternative was just as bad.

If quasars are fairly ordinary stars, then they must be fairly close by, and Hubble's redshift law does not work for them. And if it does not work for quasars, is it sensible to believe any longer that it works for galaxies? For a time, it looked as if the very foundation of observational cosmology might be in error. Some astronomers argued that quasars were 'local' objects shot out from the centre of our galaxy in a violent explosion millions of years ago, so that they could show a high Doppler redshift quite independently of the expansion of the universe as a whole. Others argued that the redshifts of quasar spectra might not be produced by the Doppler effect at all, but that these objects may be very dense, so that their light is affected by the gravitational field of the quasar—the gravitational redshift predicted by Einstein's theory of relativity.

There are still supporters to be found for these ideas, and for other, more peculiar theories. But, by and large, astronomers are now agreed on a broad explanation of quasars which not only leaves Hubble's law inviolate, but fits these mysterious objects happily into the overall scheme of the universe. Out of the apparent chaos which the discovery of quasars caused has come a stronger and better picture of the universe, which seems a more exciting and violent place than was previously dreamed.

As well as the universal violence indicated by the existence of quasars, individual members of the family can be chaotically violent themselves. If it were just a question of measuring one unique redshift in the spectrum of a quasar, the cosmologists would have a straightforward problem, even if they might still argue about the implications of its solution. But in fact many quasar spectra show several slightly different redshifts—not in the *emission* lines, however, but in the *absorption* lines. The difference is important: bright emission spectral lines come from regions of hot, excited gas but dark absorption lines are produced when cool gas lies between the source of light and the observer. So in principle you might expect to see any kind of redshifted absorption lines in light from quasars, produced by absorption of their light in cool gas clouds lying at any distance between us and the quasar. But this does not now seem a likely explanation. The patterns of absorption lines seen in quasar spectra suggest that the absorbing clouds are physically associated with the quasars, and this can reveal a lot about the structure of these energetic objects.

Several dozen quasars are known which show absorption redshifts slightly less than their corresponding emission redshifts, suggesting that these quasars are surrounded by cool material, which may have been blown out by the violence of events at the quasar nucleus. Even the few absorption redshifts which are very slightly greater than the corresponding emission redshifts can be explained, if we imagine that gas blasted out from the nucleus is now falling back towards the quasar, and so is moving slightly faster than the simple cosmological velocity of the object. But what of those quasars where the absorption redshifts are *much* less than the emission redshift?

The simplest explanation of these cases, where many quite different

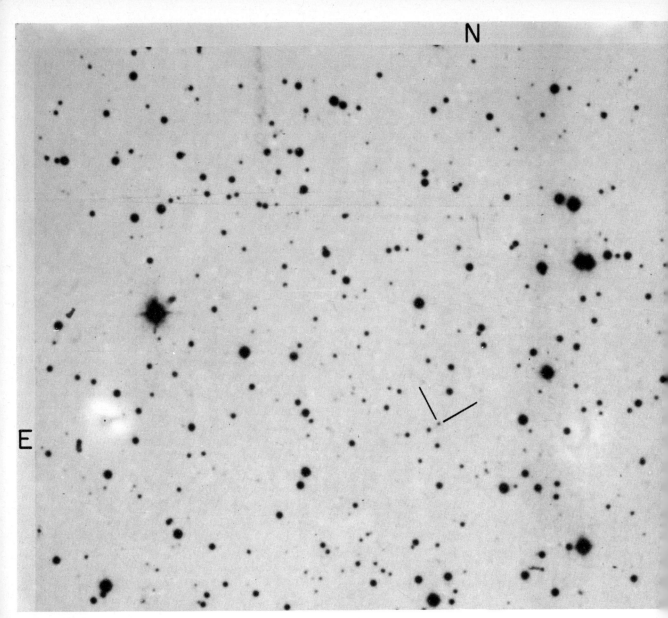

E

absorption redshifts can in some cases be found in the spectrum of one quasar, is that several shells of gas have been blasted outwards from the nucleus in successive violent outbursts. In some cases, to explain the wide difference between absorption and emission redshifts in one object, the shells must be moving outwards at half the speed of light or more—and that requires a lot of energy, even if the gas in the shell represents only a tiny fraction of the mass of the whole quasar. And this ties in very well with observations of violently energetic outbursts in other astronomical objects, helping to clarify astronomers' understanding of the relationship between quasars and other objects in the universe.

Some idea of the scope which quasars provide as a means of probing the universe is indicated by converting redshifts into velocities

Above: The radio source OH471 has been identified with the very faint quasar indicated here. Its redshift suggests that this is one of the most distant objects in the universe

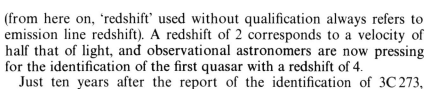

(from here on, 'redshift' used without qualification always refers to emission line redshift). A redshift of 2 corresponds to a velocity of half that of light, and observational astronomers are now pressing for the identification of the first quasar with a redshift of 4.

Just ten years after the report of the identification of 3C 273, another report in *Nature* announced that the emission redshift of the quasar OH 471 is 3·4. A couple of other quasars with redshifts this big are also known now, and the Doppler explanation of these redshifts means that these quasars are receding from us at more than 90% of the speed of light. Or, to put it another way, the light we are seeing from these quasars left them so long ago that what we are seeing now is a quasar as it was in the very early days of the universe. This is exciting enough for cosmology, as we shall see later. But in the context of the development of the new astronomy after the curtain-raising discovery of radio galaxies, the immediate interest lies in the way these objects, with their evidence of repeated energetic outbursts, can be related to other astronomical phenomena.

Throughout the 1960s, astronomers found a wealth of peculiar objects: radio galaxies like Cygnus A, in which great explosions seem to have thrown out energetic particles; other galaxies that also seem to have suffered from vast central explosions, which have left visible streams of matter pouring out from their centres; and very recently, some evidence that even our own galaxy has suffered some similar cataclysmic event in the distant past. The variety of these active galaxies, ranging from superficially ordinary systems like our own galaxy to the chaos of the galaxies known as Seyferts, and the active N-type galaxies, suggests to many astronomers that there is no clear dividing line between the categories. Rather than saying that there are many different kinds of galaxy which each do various peculiar things, they prefer one of two theories: either the activity observed in some galaxies is part of an evolutionary development, and the active galaxies are in some sense an early stage of galaxies like our own; or perhaps any galaxy might be involved in such an outburst at any time in its life, or even in a succession of such outbursts.

These ideas immediately offer an explanation of quasars. Either these objects are the first stage in the development of galaxies, or the quasars we see are just the bright, exploding central regions of whole galaxies. Such an explosion could well be so bright that the surrounding galaxy would fade into invisibility by comparison, and this idea looks one of the best now available.

In questions of cosmology and galaxy structure, it is sometimes said that there are as many theories as there are theoreticians, and this is probably true of the *detailed* theories, at least. So, as an example of the way our understanding of quasars stands at present, I shall choose one piece of work, published in 1973 by Jerome Kristian, of the Hale Observatories. Many astronomers would quibble about details of this idea, but by and large it provides a plausible account of the place of quasars in our changing universe.

Kristian started from the similarity between the active Seyfert and

N galaxies and quasars. These objects all have similar kinds of spectra; they are all the same sort of colour; they all vary in much the same kind of way, and they all show a wide spread of activity— there is no such thing as a 'typical' Seyfert, N galaxy or quasar. Kristian's suggestion that quasars are the 'egg' from which comes the 'chick' of a Seyfert galaxy which develops into the 'hen' of an N galaxy, does not seem completely satisfactory—there is no evidence that the quasars themselves are evolving from a state of 'more quasar-like' to a state of 'more Seyfert-like'. This is certainly a point which might be quibbled over, but it is quite plausible to follow along with Kristian's view that quasars are actually violent events going on at the centres of galaxies. Seyfert and N galaxies are clear examples of such activity, and all that is needed is that the quasar event should be bright enough to mask the light of the surrounding galaxy and produce a starlike image on a photographic plate. This is the idea that Kristian set out to test, using photographs of quasars which he hoped might show evidence of galaxies surrounding them. His results are impressive.

The problem with this kind of study is that a bright object, like a quasar, burns out an area of the photographic plate larger than the real image of the object. A bright point of light shows up as a small dot. The size of a quasar image depends only on its brightness, because it is a point source. But the image of a galaxy depends both on its brightness and on the size of the galaxy divided by its distance. So some cunning is needed in choosing which quasars to examine for traces of a surrounding galaxy.

Most quasar identifications are made from the astronomers' standard atlas, the *Palomar Sky Survey*; this is a set of photographs obtained with a 48-inch Schmidt telescope. But Kristian used photographs obtained with the giant 200-inch telescope, which reveal fainter objects. This is only possible after the sources have been noted as interesting, because the 200-inch is too valuable a research tool to be used for general sky survey work; so Kristian's method is very much dependent on the original Schmidt photographs, and in no way replaces it.

It turns out that quasars do appear to be associated with underlying galaxies, when they are looked at with a powerful enough telescope. Only 26 quasars had been photographed by the 200-inch at the time of Kristian's study, but all of these that should have shown evidence of an underlying galaxy, did so. Equally significantly, fourteen of the quasars studied should, according to Hubble's redshift law, be too far away for any galaxy to be visible even if one is there, and Kristian found no evidence of a galaxy associated with any of these fourteen quasars. That kind of negative evidence is very important in confirming the idea. Kristian concluded that the observations he had made were consistent with the idea that *all* quasars occur in the nuclei of giant galaxies.

Some of the very latest evidence, published only in 1974, adds considerable weight to Kristian's conclusion. After several years of intensive study, astronomers working with the 200-inch telescope found that a peculiar object, code-named BL Lac, seems to be embedded at the centre of a fairly normal giant galaxy. Since its discovery in the mid 1960s, BL Lac had disconcerted astronomers. Although it looked just like a quasar on photographic plates, it had a featureless spectrum with no bright or dark lines; so its redshift could not be measured—indeed, for all anyone could tell it might have been a blue-shifted quasar, moving *towards* us at very high speed. That would really have upset the cosmological applecart—current thinking seemed to explain quasar redshifts in the expanding universe so neatly. But now the mystery is resolved. By covering up the image of the bright central object, BL Lac itself, the team working in California were able to get photographs of the fainter galaxy surrounding it. These show a small redshift, from which the distance to the galaxy and BL Lac can be worked out.

It turns out, from the brightness of BL Lac and this redshift measurement, that the object is emitting about the same amount of energy as distant quasars, if they are really at the distances indicated by their redshifts. That is a great deal of energy, and it is still not clear exactly where it all comes from. But if nearby BL Lac is so energetic, then there is no reason why other quasars cannot be as well. Equally, if BL Lac is at the centre of a galaxy, there is still no reason to doubt Kristian's claim that all quasars are in the nuclei of giant galaxies.

And that seems the best theory at present. There is still an enormous amount of work to be done before it can be claimed that quasars are understood. What causes these galaxies to explode, and where does the energy to power them come from? There are still arguments that some or all of the redshifts might not obey Hubble's law, and these cannot yet be completely dismissed. And it is still on the cards that some as yet unthought-of theory will be conceived to explain these objects better than any idea so far. But as things stand today, at the beginning of the second decade after the discovery of 3C 273, it seems that quasars have come in from the universal cold. They are not, after all, maverick objects which will destroy our understanding of the universe, but rather another manifestation of the activity and variability of galaxies. Indeed, it may well be that the radio galaxies discovered since the 1950s are the remnants of quasar activity. The brightest and most distant quasars provide cosmologists with probes extending to the very fringes of the observable universe, and ever-improving observations of quasars and the apparently related active galaxies show just how much variety and activity there is in the universe. By the mid 1960s, astronomers in general were perhaps more prepared than ever before to find that their ideas about our changing universe might have to be drastically revised in the light of the new astronomy. But even so no one was prepared for the next shattering discovery—not some enigmatic object at the fringes of the universe, but an unprecedented phenomenon in our own astronomical backyard, our own galaxy.

4 Pulsars

For twenty years, all the excitement in astronomy seemed to be related to events outside our galaxy. Cosmology, radio galaxies and quasars dominated the astronomical headlines, and by the mid 1960s there was a general feeling that the study of stars in our own galaxy was a routine operation, rather boring and insipid compared to the discoveries made beyond it, as improving techniques enabled astronomers to push their focus of vision ever further into the depths of the universe. That was certainly how it seemed to me as a brash student in 1966 and 1967; but I was wrong, along with many other astronomers. Since 1968 a succession of 'local' discoveries has shown just how exciting our own galaxy is—and a better understanding of the remarkable phenomena uncovered seems likely to lead to a better understanding of the workings of galaxies in general. The first of these surprises to astonish the astronomical community was, of course, the discovery of the remarkable radio stars known as pulsars.

The discovery itself was another example of scientific serendipity—no one suspected that rapidly pulsing radio sources might exist, and when the signals from them were first found they were taken to be interference.

In July 1967, a large radio telescope at Cambridge began operations intended to reveal more about the detailed structure of radio sources. This particular instrument used a technique which we have not yet mentioned—the so-called 'scintillation' of small radio sources, when viewed through the tenuous gas of interplanetary space. This scintillation is exactly equivalent to the twinkling of ordinary optical stars when viewed through the Earth's atmosphere. Planets, of course, do not twinkle, and that is because their images are too big; in just the same way, radio astronomers can tell that any radio source which twinkles, or scintillates, must be smaller (in terms of the degree of arc covered) than a critical size, which depends on details of the gas of interplanetary space.

So the Cambridge group built an instrument specially designed to detect rapid flickering of radio sources. The group was far from pleased when their new instrument seemed to be plagued by regular interference. It is an occupational hazard of radio astronomy that man's growing use of radio bands, and more recently the presence of artificial satellites broadcasting even from above the atmosphere, make it difficult to observe some kinds of radio source. But within a couple of months it was clear to the Cambridge researchers that the signals were in fact not man-made at all. Instead, they were regular pulses of radio noise, in bursts about a third of a second long, separated by periods of about one and a third seconds, and coming from a fixed point in space. That caused a great deal of excitement—and even consternation. The radio pulses were so regular, like the ticking of an interstellar clock, that the possibility had to be seriously considered that they might be signals from intelligent beings on a planet circling another star. Not entirely facetiously, the first pulsar was soon being referred to around Cambridge as 'LGM 1', with the initials standing for 'Little Green Man'!

This idea had to be discarded when more observations revealed that the signals showed none of the variations to be expected if they came from a planet in orbit around a star, and that they must be coming from some kind of optically invisible star itself. As more pulsars were discovered—and once the news was out many radio observatories found it easy enough to detect pulsars—it became quite clear that they must be a natural phenomenon. The explanation of that natural phenomenon was, however, almost as exciting for the new astronomy as the discovery of real signals from another civilisation would have been.

The first pulsars all had 'periods' (regular intervals between pulses) of a few seconds or so—and their periods were regular to an accuracy as precise as 0·000002 sec. What could produce such a regular burst of radio noise? In astronomy, when you find a regular variation there are two main ways of explaining it: either something must be rotating, or something must be pulsating. The rotation explanation can be broken down into two parts; either we could be seeing a 'spot' of radio emission on a single spinning star, or we could be viewing a 'binary system' in which the radio star is periodically eclipsed by its companion.

Throughout 1968 virtually every theoretical astronomer in the world had a go at explaining pulsars by one or other of these basic models, with the theories being modified and, in all but one case, eventually thrown out, as new pulsars were discovered with new properties to be explained.

The first theory to fall by the wayside (in fact, it never really got off the ground) was the binary theory. It is just too difficult to invent a binary system which would be stable enough to produce such regular 'ticking', and where the radio signal could be flashed on and off quickly enough to explain the very clear-cut pulses seen from pulsars.

Of the other two theories, the pulsating model was the front runner at first. That was particularly exciting for me, because in 1968 I was a research student in Cambridge, and since I had already been studying the theory of stellar pulsations I was able to change the direction of my work slightly and apply it to pulsar models without difficulty. At the time, there was a great friendly rivalry among astronomers around the world working on similar theories. But everyone was thoroughly perplexed because the rate of pulsation or 'periods' of the first pulsars measured fell exactly into a gap which could not be accounted for by any of the standard models of different kinds of stars.

Such standard models are derived by getting together all the equations which describe the inside of a star—equations relating pressure, density and temperature—and using an electronic computer to calculate a mathematical model in which all the equations are nicely in balance. Once you have a balanced, stable model it is possible to find out how it will pulsate: one or two of the numbers in the equations are slightly changed and the computer calculates how all the other numbers must vary as a result.

The only difficulty with this is the number of equations and variables (pressure and so on) which must be included. Even the best computer models can only be an approximation to the real thing; although they can provide broad outlines of behaviour for different types of stars. And those broad outlines just did not fit pulsars at all.

It's not always true that everyday commonsense ideas can be applied to stars; but one such idea can, and that is that the bigger a star is the slower it pulsates. The smallest stars of all are neutron stars; in early 1968, all the computer calculations of model neutron stars showed that they could be expected to oscillate with periods of a few thousandths or hundredths of a second—far too rapidly to explain pulsars. The next size up in stars are the so-called white dwarfs; but according to the model calculations, these vibrated rather too slowly to explain the known pulsar periods. It just did not seem possible to make a mathematical model of a star which would pulsate every second or so; that is, with a period close to the pulsar periods.

But the computer modelling technique leaves plenty of scope for human error, and it is one of the tasks (almost a duty!) of research students to do this kind of work again in a slightly different way, to see if the generally accepted models really are correct.

In the case of white dwarf pulsations, it turned out that the models which everyone had been using as the basis of their calculations were faulty. The error was very small, and only applied to certain models of very dense stars—but it was just enough. From work which I did with Dr John Faulkner, it became clear that the standard white dwarf models had been 'made' using a very slightly incorrect assumption about some of the most critical numbers which have to be fed in to the computers; with these corrected, we were able to 'construct' models of white dwarf stars that would oscillate quite happily with longer periods close to those of the pulsars known at the time. And if we included the more complicated effects that might arise in a rotating star, it seemed that oscillation periods even shorter than one second could be found.

Our moment of glory was, however, short-lived. No sooner had we made this discovery, it seemed, than the radio astronomers found a whole clutch of pulsars with much quicker pulses, going right down to a few thousandths of a second. There was nothing wrong with our calculations (as far as I know), and the pulsar excitement had certainly made us clear up an error in astronomers' understanding of white dwarf stars. But there was no plausible way in which white dwarf pulsations could account for the rapidly flashing signals from the pulsars being discovered by the end of 1968.

That left neutron stars. They certainly could not pulsate slowly enough to explain the slowest pulsars, but they could rotate at all kinds of speeds. Today, there are very few astronomers who do not believe that pulsars are rotating neutron stars, which beam out a radio pulse every time they rotate, rather like radio 'lighthouses' in space. The clincher which converted everyone to this theory was the discovery that most pulsar signals are slowing down by a tiny amount, just as they should do if their source is a spinning star that loses energy and momentum.

In many ways, this was the most exciting of all the pulsar models. Astronomers already knew about white dwarfs from studies of visible stars, and although they are interesting enough there would have been nothing fundamentally exciting in the discovery of more of them. But neutron stars had never been detected before, even though their existence had been predicted from theory.

The very broad outline of the theory of stellar evolution runs something as follows: When a star forms from a collapsing cloud of gas, it gets hot in the middle; because, as the cloud contracts, its gravitational (potential) energy gets converted into heat (kinetic) energy of the hydrogen gas molecules in the cloud. When a high enough temperature is reached, nuclear fusion reactions begin in the middle of the star, turning hydrogen into helium and releasing the heat energy needed to keep a stable star going. Eventually, all the hydrogen is used up (eventually being something like ten thousand million years, so there's no need to worry about it happening to our Sun just yet), and the star contracts a bit more, until the temperature is high enough for nuclear fusion of helium into still more massive elements. Although the middle of the star is now denser, it is hotter too; so the gas 'atmosphere' of the star expands, and it can become a giant star as big as the whole inner part of our Solar System.

Then life begins to get interesting. When the star has used up its helium and no more nuclear reactions can occur, it either fades away, or goes out with a bang—literally. The stars which fade away just cool down and after a few million years become a more or less solid lump—one giant crystal, perhaps as massive as the Sun but with a radius more like that of the Earth. And there they will sit forever (barring one of the more bizarre happenings mentioned in Chapter 10). But what of the stars which go bang—the novas and supernovas?

In the case of supernovas, at least, the theory said that they might very well produce a neutron star remnant. The reasons why supernovas go bang are still not completely clear, but it looks rather as if in some stars the nuclear reactions run riot when the conditions are just right. This leads to a huge explosion, which both blasts material outwards from the surface of the star to form a spectacular remnant, and also squeezes the innermost part of the star as tight as it will go. And that, according to the best theories, is tight enough to form a whole new kind of star—the neutron star.

In very general terms, a neutron star is one that has been compressed so much that all its electrons and positrons have combined into neutrons. There are no more atoms in such a star—just a sea of neutrons, rather like one giant atomic nucleus, but without any protons. Such stars might well end up (after the explosion in which they were formed) with as much mass as the Sun has today; but if so, it would all be packed into a radius of only about ten kilometres.

Now that is a pretty outrageously extreme kind of star. Certainly the best theories of physics predicted that neutron stars might exist, but not everyone was sure that the theories really held good to such extremes.

So the discovery that pulsars might very well be neutron stars stirred up even more interest in these remarkable objects. The next question was: are any of the pulsars associated with supernova remnants, where neutron stars ought to be found? And the rapid discovery of a positive answer, in the form of a pulsar in the best known supernova remnant of all, the Crab Nebula, converted most of the remaining doubters. It seems that neutron stars do exist, and in just the spot that theory predicts.

5 The Crab Nebula

The Crab Nebula, in its relatively brief life over the past 900 years, has surprised astronomers with a series of unexpected discoveries. The first surprise came in June AD 1054, when Chinese and Japanese astronomers noted the sudden appearance of a new star, bright enough for a whole galaxy. (Some people argue that this 'guest star' was not actually associated with what we now call the Crab Nebula, but few astronomers take this seriously.) In the present century, the connection between this ancient supernova and the nebulous gas cloud now visible was realised by the great astronomer Edwin Hubble. His measurements of the rate of expansion of the cloud (using the Doppler effect) showed it to be blasting outwards at about 1000 km a second. It can easily be shown that the kind of cloud we see today could have been produced by such expansion, provided it started from a stellar explosion about 900 years ago. But for many years it was difficult to explain the nature of the powerhouse that still keeps the cloud expanding.

Then, in 1949 the Crab Nebula was the first astronomical radio source to be definitely identified with an optical object (apart, of course, from the Sun). A few years later the observers noticed ripples of activity in the wisps of cloud which make up the Nebula, showing that it is still very much an active region, with something at the centre stirring it up. In 1963, the new discovery was that there are X-rays coming from the centre of the Crab Nebula, and in 1968 the source of all this energy was at last unveiled, when the pulsar, NP 0532, was found at the centre of the Crab.

It is hardly surprising that astronomers often remark, at least semi-seriously, that observational astronomy can be divided into two parts: the study of the Crab Nebula, and the study of everything else. For in the Crab they can see examples of most of the interesting objects in our galaxy, gathered in one place and interacting in complex ways. But the key to understanding this complex behaviour is definitely the Crab pulsar itself.

The firm evidence that pulsars are composed of neutron matter—the most compressed form of matter possible—rested upon the observations of the Crab, and it is difficult to recall today, just five years later, the excitement that those observations caused. The crucial point was that the Crab pulsar was found to have a radio 'period' of only 33·1 milliseconds; and this rate of pulsation is just too quick to explain in any sensible way except as the rotation of a tiny object which is at the same time a powerful source of energy—in other words, a neutron star. Together with this discovery came

Right: The Veil Nebula—
tenuous gases blasting
outwards from the site of a
supernova explosion

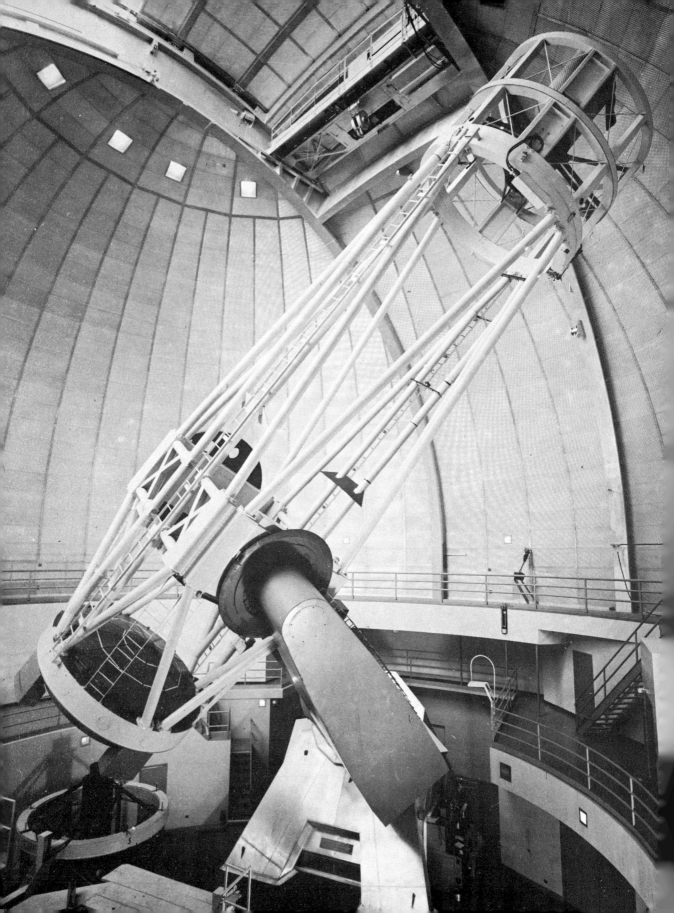

the observation that the pulsar is slowing down very gradually, and this is just what ought to happen as the spinning star gets older and loses energy and momentum. Other pulsars are now also known to be slowing down in just the same way, and all the evidence suggests that the oldest pulsars are the slowest, again just as the theory predicts. So that accounts for the discovery of such a fast pulsar in such a very young supernova remnant—young, that is, in astronomical terms.

Young pulsars should be the most active in many ways, and astronomers were particularly delighted in 1969 by the discovery of optical flashes from the Crab pulsar. In fact, the flashing star, which we now know to be the pulsar, had long been known and is easy enough to observe. But, of course, no one dreamed of looking for flashes with a period of 33 milliseconds, and any kind of photograph at all simply gave an average of hundreds or thousands of flashes, so that the star looked like a normal object with constant brightness.

So how did observers measure the flashes? They were only able to do this, in fact, because they knew exactly what they were looking for. The radio signals had already pinpointed the position of the star itself and the period of the flash that it should be producing. The team at the Lick Observatory who made the discovery used a mechanical rotating shutter attached to their 120-inch telescope, to 'chop' the light from the suspected pulsar. The shutter was simply a disc with half a dozen slots cut in it, which could be spun at various speeds; at one adjustment of the shutter, the slots allowed light through just when the pulsar was 'on', and at others just when it was 'off'. So now photographs taken through the shutter/telescope system could produce average pictures corresponding to each of these states—but only if the target star really was flashing at exactly the radio pulse frequency. So the Lick team entered the history books, the Crab Nebula became even more famous and received even more intensive study.

One of the oddest results of this attention came when a team of X-ray astronomers at Rice University, Texas, decided to take another look at some tapes of the information they got on an X-ray-experiment balloon flight in 1967. X-rays from astronomical sources can only be detected by flying the detectors above most of the atmosphere in balloons, rockets or satellites. This is because X-rays are blocked out by the atmosphere, which is just as well for us or we would be burnt to a turn. We will learn more of the exciting new science of X-ray astronomy later on (Chapters 8 and 9); the interest from the point of view of students of the Crab Nebula is that X-rays only come from very energetic sources, and the Crab is certainly one of those.

The Rice team had found X-rays from the Crab all right, on their June 1967 balloon flight—but in 1967 nobody had any notion that anything in the universe, let alone in the Crab, might be pulsing 30 times a second. It was only in 1969, after the discovery of radio and optical pulses from NP 0532, that the Rice tapes were

re-examined to see if a 33-millisecond period was present in the X-ray data. Sure enough, it was. Once again you can see how important serendipity can be to the astronomer! How galling it must have been for these astronomers to know that, if there had been any reason for them to look for such an effect, they could have found the first X-ray pulsar *before* anyone found any radio pulsars! There is certainly a lesson here, and one which astronomers have learnt well— never throw your old data or recordings away, you don't know when they might come in useful.

But these June 1967 tapes were more than a curiosity. They showed how the period of NP0532 had changed between 1967 and 1969, and this helped to improve the theories which were beginning to be produced to account for the slowing down of the Crab pulsar (and of other pulsars). Those improved theories had soon, however, to be rather extensively revised. By the end of 1969, astronomers knew that the slowing-down was not steady after all, but that the pulsar could suddenly speed up slightly before settling into its usual slowing-down pattern again.

Astronomers like inventing new technical terms, and these sudden events soon became known as 'glitches' or, more prosaically, 'spinups'. Several bright ideas were offered to account for this behaviour. One which seemed interesting, but which does not seem to be generally accepted now, is that a planet in orbit around the pulsar is producing a regular shaking-up of the spinning neutron star. Other theoreticians argue, more successfully, that the spinup could be explained by star-quakes—the solid crust of the neutron star settling down into a new pattern.

That sounds remarkable enough when said quickly, but the theory really does seem to work. The idea is that a rapidly spinning star must be distorted from a sphere, bulging heavily at the equator. As the spin slows down, the star would naturally tend to become more spherical—but since it has a solid crust, it cannot change shape. Eventually, so much momentum is lost that something has to give; the crust cracks and is forced into a more spherical position, and so the process goes on. One of the nicest things about this theory is that it can also explain how the pulsar stirs up the central wisps in the nebula surrounding the pulsar. Some observations indicate that these wisps are pushed around just after a spinup, and indeed energy should be released in a spinup event.

What about the energy still pushing the Crab Nebula outwards at 1000 km a second, 900 years after the original supernova explosion? That too is explained by the presence of the pulsar. NP0532 is pushing out intense beams of energy right across the electromagnetic spectrum, from radio waves through optical light to X-rays, and the amount of this energy seems, from the observations of the pulsar, to agree very well with the energy requirements of the driving force of the Nebula. The question still to be answered is: how do the pulsars pulse?

It is something of an embarrassment to astronomers today that, after six years of intensive study, they still cannot really answer that question properly. It's easy enough to make vague explanations, and the most popular idea is that the pulsing is produced by an effect of the rapidly spinning magnetic field of the pulsar. If a star starts out, like the Sun, with a fairly ordinary magnetic field then, by the time it has been through a supernova explosion and squeezed down to a diameter of less than 50 km, the magnetic field has been pretty drastically squeezed too. Squeezing a magnetic field, so that the lines of force are closer together, makes the field stronger; so every pulsar probably has a strong magnetic field being swept around with it.

Now this leads to some interesting effects. In particular, pulsars spin so quickly that somewhere not too far above the pulsar's surface the magnetic lines of force will be being swept around at the speed of light—and Einstein's theory of relativity tells us that odd things happen at that kind of speed. The odd things don't happen so much to the field itself, but to any charged particles—protons, electrons and so on—which it sweeps up.

These particles could very well be trickling out from the pulsar, at each magnetic pole, and following the magnetic lines of force upwards and outwards. As they do so, they spin faster and faster; and since an accelerated charged particle emits radiation, they radiate. They radiate in a particularly spectacular fashion out where the spin approaches the speed of light—on the 'speed of light circle'. That radiation is probably what we observe as the flashing of a pulsar at radio, optical or X-ray frequencies. But just how the exact type of pulse astronomers can detect is produced, nobody knows.

There is certainly plenty of scope for more astronomers to try their hands at explaining details of the pulsar mechanism, in particular what goes on between the surface of the pulsar and the 'speed of light circle'. But the story does not stop there. As we have already seen, the stream of energy and particles pouring out from a pulsar can power the activity of an entire nebula such as the Crab. On the scale of the whole galaxy, however, even that fascinating source is just a tiny detail. What goes on beyond the Crab Nebula itself?

A few years ago, most astronomers would have said that hardly anything went on in interstellar space. But all those accelerated particles have got to go somewhere—indeed the whole Crab Nebula is expanding and disappearing, slowly by human standards but at a frantic pace by the standards of the galaxy. Today, it looks as if the accelerated particles become cosmic rays, some of which reach the Earth—to tell us more about the nature of the universe than any particles accelerated in man-made 'atom smashers'. Just what they can tell us, we shall see in Chapter 7; but their more leisurely counterparts, the molecules of the Crab Nebula drifting off gently at 1000 km a second, also have an interesting future. Indeed, molecules which started out in much the same way could have been the precursors of life.

6 Molecules in Space

All the atoms of which we are made, except possibly the hydrogen ones, have been processed inside stars—'cooked' by nuclear burning, and built up from hydrogen, the basic atom of the primeval universe. The carbon in the paper of this book, the atoms of the clothes we wear and of the food we eat are all the product of this nuclear burning, as are the atoms of our own bodies. In a very real and literal sense, we are all 'stardust'. And this stardust got to where it is today as a result of explosions like the one which produced the Crab Nebula. This was all known some time ago: astronomers realised that this was how all the heavy elements must have been made and dispersed, long before radio astronomers began to unlock the secrets of the clouds of cool dust and gas which are scattered throughout our galaxy.

Generally, it used to be taken for granted that while atoms might be produced and scattered in this way, the next step in the production of conditions like those on Earth, and eventually the beginnings of life itself, must take place on planets. That step, of course, is the build-up of molecules, and particularly organic molecules, from atomic combinations—chemical reactions—involving various elements.

In the 1930s, a few simple compounds were identified; their characteristic absorption lines showed up superimposed on the spectra of hot stars. These diatomic molecules were CH, CH^+ and CN; in 1963 another molecule, OH, was found by astronomers using a radio telescope. All these molecules were present in cool clouds in interstellar space, and were detected because, when light or radio waves shine through those clouds, the molecules absorb radiation at certain characteristic frequencies, leaving dark lines in optical spectra, or dips in traces of radio signals. But those lines and dips corresponded to just about the simplest molecules possible, and it was neither surprising nor of more than passing interest to most astronomers that such molecules could be found in interstellar space. Then, in 1968 (the same year that pulsars were discovered), the situation was changed dramatically.

The change seems to have been stimulated when Professor C. H. Townes, a Nobel Prize-winning physicist, became interested in astronomy—in particular, in radio-astronomy at microwave frequencies. These wavelengths, a few centimetres or so, are just the ones where many fairly complex molecules produce characteristic spectral lines; but they require accurately built radio telescopes, with smooth surfaces. The more interesting parts of the spectrum, from the point

Right: 30 Doradus—a nebula in the Large Magellanic Cloud. This nebulosity shines because of the energy released by the hot young stars inside it

44

of view of discovering molecules in space, seem to be at fairly short wavelengths, by radio astronomy standards; and the shorter the wavelength, the more accurately the telescope must be built. This is obvious enough: if you are measuring something that is about a metre long, then whether it's a wavelength or a loaf of bread, an error of 1 cm is not too important. But if you are measuring something about a centimetre long, then to get the same relative accuracy you need measuring instruments accurate to 1 mm.

There is also another snag about measuring and detecting the radio spectra of complex molecules (or 'polyatomic' molecules)—you have to know what you are looking for. But many of the interesting molecules have not been studied thoroughly in the laboratory yet, so physicists and astronomers just don't know for sure what spectral lines to expect. So the practical approach is just the opposite of that used when looking for elements in a star. Then, astronomers simply photograph the spectrum of the star, measure the positions of the characteristic lines, and look up which elements they correspond to. With radio spectra of molecules in space, however, you first pick a likely molecule, then decide on a likely feature to look out for, corresponding to a particular 'transition' of the molecule, and only then do you actually switch your radio telescope to that frequency and start looking.

That is what a Berkeley team, inspired by Professor Townes, did in 1968; and they hit the astronomical jackpot (no serendipity here, just foresight and careful planning). In December of that year came the announcement that ammonia (NH_3) had been found by its radio emission at a characteristic wavelength of 1·26 cm. Soon after this, the same team found radio emission from water molecules in space; and then another group of American astronomers reported that they had detected radio traces which could best be explained as coming from formaldehyde. That really put the cat among the pigeons, because formaldehyde is not just a more complex molecule (H_2CO); it is an organic molecule which could only have been built up from other compounds such as methane in the cool interstellar clouds. Even at that time, early in 1969, the more adventurous astronomical minds were tempted to speculate that there might be a kind of molecular soup in these clouds, with many organic molecules, and perhaps including such complex varieties as amino acids, essential prerequisites for life as we know it on Earth.

The interest aroused by these discoveries stirred many radio astronomy groups into joining in the hunt for interstellar molecules, and an impressive list was quickly racked up. Carbon monoxide was the next, and within a couple of years such exotica as formic acid and cyanoacetylene were also found. More than 25 polyatomic molecules have now been found in interstellar space, and of these all but a couple contain carbon. For some reason, the production of organic (carbon-containing) molecules seems to be favoured in the clouds of gas and dust which exist between the stars of our galaxy. The reason for this is to some extent a mystery, but it raises

intriguing possibilities regarding the origin of life, and the nature of life. In 1974 one of these molecules, formaldehyde, was detected in two other galaxies, and this makes the possibilities even more intriguing.

The importance of carbon in building up these complex molecules is probably due to a combination of two effects. First, it bonds very strongly to other atoms, and so molecules containing carbon are less likely than other molecules to be knocked apart by energetic particles such as cosmic rays (of which more later). Second, carbon has four 'active bonds'—it can bond with as many as four other atoms at a time—and this helps it to build up many families of molecules based on groupings like CN, CS, CC and CO. The most puzzling question, however, is not how the molecules stay together once they are formed, but how the atoms first get together.

The answer to that seems to lie in the presence of tiny dust grains in interstellar space. These grains are believed to be made up of a small core of solid carbon, about one ten-thousandth of a millimetre in diameter, covered by an icy layer of organic molecules which have built up on the core one atom at a time. Once an atom has stuck to the carbon, other atoms which come along can also stick on and combine to make molecules. The molecules can then get out into space again as a gas if cosmic rays with sufficient energy collide with the ice-covered dust grain. All that is a pretty sketchy theory, and certainly there are plenty of details for the new astronomers to be working on. But it is what happens later on that is really intriguing, after the clouds, which we now know do contain many complex organic molecules, evolve to the next stage of their development.

These cool clouds in interstellar space are just temporary features in the life of the galaxy. They soon collapse, under their own gravity, and form new stars. That is certainly where our Sun started, because our Sun contains quite a lot of heavy elements—it is a second generation star, made out of atoms which have already been cooked in other stars and expelled in great explosions. Again, what exactly happens when a cloud of dust, atoms and molecules collapses is far from clear in detail. But we do know for sure that it is possible for the cloud to condense, not just into one object—a star— but into a star plus a family of planets.

That must be so, or we wouldn't be sitting here on a planet thinking about the problem. And that means, according to the new microwave discoveries, that right from the start the planets in such a system, and the planets in our own Solar System, must have a liberal sprinkling of organic molecules in them. This certainly means molecules like H_2CNH, $HCCCN$, H_3CCOH and others which have been discovered in space already; it may mean still more elaborate compounds such as amino acids.

This puts a whole new complexion on theories of the origin of life. More than twenty years ago, experiments were carried out in which water, ammonia and methane were mixed in a bottle and activated by an electric spark. The molecules combined to produce amino acids, and biologists and chemists were convinced that provided the methane, ammonia and water were present on Earth in the distant past, then lightning and Nature could have done the rest, building up ever more complex molecules, and eventually producing life. But the interstellar soup contributes even more complex molecules as 'starters'. Formaldehyde, for example, is a forerunner (in the 'chemical evolution' sense) of sugars; and cyanoacetylene leads to the build-up of molecules which in turn are essential to producing DNA, the 'life molecule' itself.

All this ties in neatly with other recent discoveries. Some pieces of meteorites which have survived the passage through the atmosphere to land on the surface of the Earth have been examined and found to contain such complex molecules as amino acids; it is a difficult and subtle question to sort out whether these acids 'belong' to the meteorites, or whether they have been 'contaminated' by handling on Earth. But it certainly looks as if the molecules do come from outside our planet. The link with molecules in interstellar clouds looks fairly clear. Meteorites and, in particular, comets must contain a lot of leftover molecules from the formation of the Solar System. Indeed, you could say that comets are in a sense 'fossil' remnants of the interstellar cloud from which we formed. So one possible course of events is that after the Earth formed and was reasonably cool, collisions with comets and meteorites, which must have been much more common just after the Solar System formed, scattered the precursor molecules needed for life to form throughout the atmosphere and oceans. Another possibility is that the molecules were indeed built up on the Earth itself. But whatever the details of the process, it seems a fair bet that the chemical origins of life are present at the birth of a planet, and that evolution can begin then—if, indeed, it has not already begun in interstellar space.

There are two implications which together make these discoveries some of the most important ever made, not just in astronomy but in the entire history of mankind. In the first place, it looks very much as if the process in which an interstellar cloud gives birth to a solar system *automatically* leads to conditions on the planets of the system which are suitable, in some cases, for life to develop. So we may very well not be alone in our galaxy. Second, the kind of organic molecules essential for life on Earth seem to be just the same as the organic molecules we can detect in space. We have seen that carbon seems to be the key to the build-up of complex molecules, and why this should be so. And that means that chemical evolution, eventually leading to biological evolution, must start on any new planet from the same kind of molecules which started the evolutionary process on Earth. Of course, evolution can take many curious branches, as the diversity of life on Earth testifies. But it looks very much as if any life 'out there' is not the kind of silicon-based, ammonia-breathing monster familiar from certain kinds of science fiction, but is rather like us, at a chemical level at least.

Overleaf: Dark lines of cold interstellar material known as 'Elephant Trunks' seen mixing with the hot gas surrounding a newly formed star

49

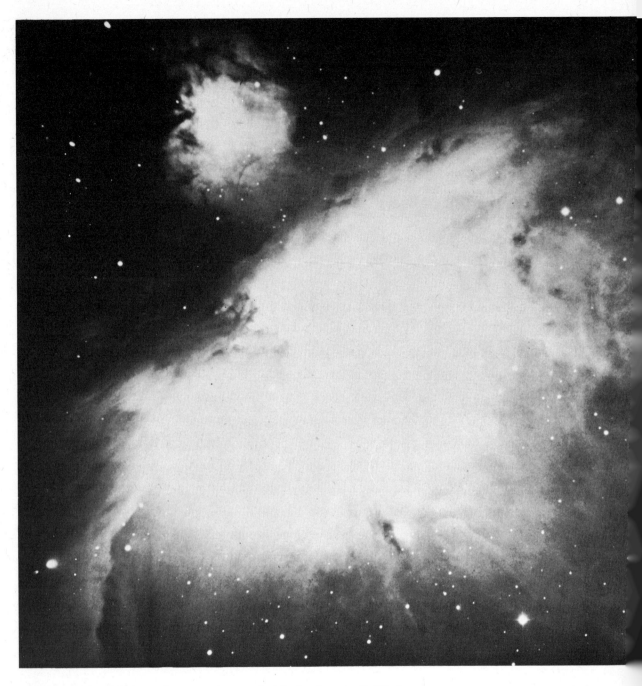

Amino acids, DNA and so on are probably basic to all life in our galaxy, which makes us brothers under the skin, whatever outward forms other creatures may have.

We may never know for sure that there is other intelligent life in the universe, but the odds now seem to favour it. Far from our own planet being unique in this respect, the discovery of molecules like formaldehyde in other galaxies suggests that even our own galaxy

Two views of the Great Nebula in Orion: photographed (left) using an optical telescope and mapped (right) by a radio telescope at 110GHz

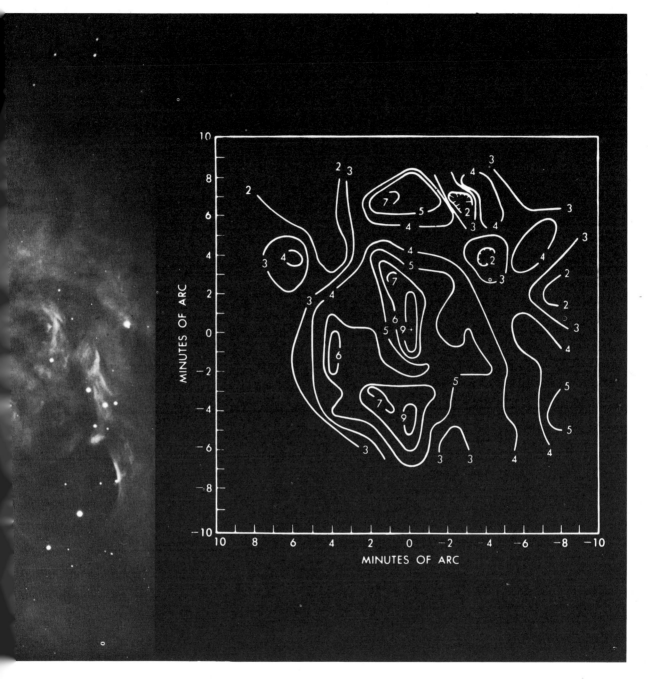

may not be unusual in possessing planets on which intelligent beings live. It is a curious and, to my mind, rather comforting, thought that somewhere in the galaxy NGC253 there may be carbon-based lifeforms with DNA in their cells, studying the heavens with their radio telescopes, and that they may be speculating on the significance of the detection of formaldehyde emission from the rather unspectacular collection of stars which it pleases us to call 'our' galaxy.

7 Cosmic Rays –Tachyons & Quarks?

We have seen how the particles which are visible in supernova remnants like the Crab Nebula can develop, together with other particles of dust and gas in interstellar space, into molecules with a fair degree of complexity, and how these molecules might play a part in the development of life on planets. But what of the other particles produced by supernovae and neutron stars—the high-energy particles sleeting out from these energetic sources at speeds close to the speed of light, thirty thousand million centimetres a second? These particles contribute, in no small measure, to another branch of the new astronomy, the study of cosmic rays.

Cosmic rays are simply charged particles—atomic nuclei stripped of their electrons—which shower onto the Earth from space. Some of the particles are simple, such as the humble proton (hydrogen nucleus) or alpha particle (helium nucleus); others are big and, relatively, heavy—nuclei of iron, for example. They can be detected by the tracks they leave in photographic emulsions, or in cloud and bubble chambers, and have been studied for decades. But the development of cosmic ray astronomy really leapt forward, like so much else, with the advent of rockets and balloons to carry the detectors above the atmosphere. That is important because it provides a means to look at the raw 'spectrum' of cosmic rays, before they have interacted with the atmosphere. Those interactions are spectacular: one highly energetic particle hitting the top of the atmosphere produces a shower of secondary charged particles which descend to be picked up on ground-based detectors, making ground-based cosmic ray astronomy a complex piece of detective work. Basically, astronomers often have to work backwards from their observation of a shower of lighter particles to calculate what large particle it was that originally hit the atmosphere, and where it was coming from. It can be done; but it is not easy, and the interpretation of a cosmic ray shower sometimes depends on who is doing the interpreting.

Such difficulties aside, however, the presence of such energetic particles in the universe offers one of the best hopes of unlocking some of nature's deepest secrets. During the 1960s, and to a lesser extent recently, we heard a lot about giant, expensive particle accelerators, or 'atom smashers'. Such machines, synchrotrons, cyclotrons and the like, are now so costly that several nations have to club together to build them. They are, in financial terms, the very biggest of big science, swallowing up millions of pounds in building costs, and comparable sums in running costs. The motivation for this vast expense lies in the hope that, by studying interactions of particles at very high energies, physicists will be able to find out the fundamental nature of these particles, and the most fundamental laws of physics. But the search seems endless: the atom itself has long been known to be divisible, and now it seems not only that the nucleus of the atom can be divided but that even protons and neutrons are divisible into more fundamental particles. The more energy the physicists put into their machines the more leptons, pions, chions and so-ons they seem to get out. And as the accelerators get bigger, more and more energy (and money) is needed to find out anything new.

Indeed, the stage has now been reached where physicists have as much of a struggle to raise funds as they do to use the funds properly; and the irony of this is that all the while the Earth is being provided, free of charge, with an abundant supply of particles far more energetic than any made by man. These, of course, are the more energetic of the cosmic rays.

Just how do cosmic ray energies compare with those obtainable in man-made accelerators? A good comparison is obtained by noting that the immediate goal of particle physicists, their Holy Grail, is the development of a 1000 GeV machine. The particle energies are measured in electron volts (eV), and 1 GeV is 1000 million eV; it need not worry us just what 1 eV is, since we are comparing two energies measured in the same units. In cosmic ray terms 1000 GeV (or 10^{12} eV) is a piffling energy; cosmic rays with energies a thousand or a million times greater are not uncommon, and some have been detected with energies well in excess of 10^{20} eV.

Two cosmic ray studies in particular have come close to two of the most fundamental of nature's mysteries. These studies involve the possible detection of 'quarks' and 'tachyons': quarks are the bizarre particles which many theoreticians believe must exist in order to explain the observations made by particle physicists from their accelerator experiments; tachyons are particles which can, even according to Einstein's theory of relativity, travel faster than light.

Before we get too involved in these particulars, it is as well to look at the nature of cosmic rays. There is still some mystery about where they all come from (perhaps that is what puts the particle physicists off?), but some are known to be produced by the Sun, many almost certainly by pulsars, and others in violent explosions of stars and even whole galaxies throughout the universe. There is a continuing debate among astronomers about whether or not cosmic rays from outside our own galaxy can ever penetrate to the Earth; perhaps all of our cosmic rays are 'local' in that they come from energetic sources of one kind or another within the galaxy. But their origin does not matter at all to the seekers after tachyons and quarks.

In the old days of physics, the process was basically one of explaining observations. Newton, we are told, observed an apple to fall and was inspired to formulate his theory of gravity. But today things are, more often than not, just the other way around. The

Above and right: Cosmic ray detectors at Durham University

theoreticians scratch their heads and puzzle out some more or less bizarre process which the theory predicts should happen, and then the experimentalists are put on the scent in the hope of detecting the predicted effect and confirming the accuracy of the prediction. Such is the case with tachyons.

In the 'standard' interpretation of relativity, we learn that no particle can ever exceed the velocity of light, because its total energy is given by the expression:

$$m_0 c^2 (1 - (v^2/c^2))^{-1/2}$$

where m_0 = mass of particle
  v = velocity of particle
  c = speed of light

When $v = c$, the energy becomes infinite, which is clearly absurd; and when v excccds c the energy is imaginary, which is even more absurd. But there is a way round this absurdity, and just the kind of way round which appeals to mathematical physicists. We can keep the total energy real and not imaginary provided that the mass m_0 is imaginary; then, if v exceeds c the energy is given by the product of two imaginary numbers, that is, a real number.

Now, you might very well say that you are no more inclined to believe in a particle with imaginary mass (*im*, say) than in one with imaginary energy. We live in a democratic country, and I would fully respect your right to stick to that argument. But if you do consider the implications of allowing particles of imaginary mass into the equations, life becomes much more interesting.

To start with, in the inverted world of the tachyon—the faster-than-light particle—energy loss corresponds to an *increase* in speed. This is just the opposite of our world (and, for mathematically-minded readers, it is a direct consequence of the appearance of the product $i \times i = -1$ in the equations). Also, any charged tachyon must constantly emit radiation in the form of light pulses, through a process known as the 'Čerenkov effect'. This has been likened (rather inaccurately but graphically) to an optical equivalent of the sonic boom produced by an aircraft travelling faster than the speed of sound. So our poor tachyon must lose energy, and must therefore travel faster and faster until all its energy is gone. In the graphic words of one mathematical astronomer, 'the theory predicts that the universe is swimming in a sea of clapped-out tachyons'.

So much for the theory; what of the practice? The physicists with their giant accelerators have not had a lot of luck detecting tachyons. Even if they could get enough energy to produce some from a collision between other energetic particles, they would be in a lot of trouble when it came to measuring their existence before they accelerated off into the sunset. But the cosmic ray physicists have done much better—or at least, some of them think they might have detected tachyons.

The key observation depends on measurements of the shower of secondary cosmic rays produced when an energetic particle arrives at the top of the atmosphere. If any tachyons are present in this shower, then they must, by definition, speed on ahead and reach the ground rather sooner than the ordinary slower-than-light cosmic rays. The atmosphere is just thick enough that this delay might be noticeable, in the form of a response by ground-based detectors a few tens of microseconds before the shower proper arrives. And that is just what a group of Australian physicists think they have observed. In cautious jargon they claimed only that they 'observed non-random events preceding the arrival of an extensive air shower'. That undramatic claim has, however, thrown the cosmic ray community into a flurry of activity. Perhaps by the time you read this, tachyons will have been detected by other groups, or perhaps there will be an alternative explanation of the observations, but one thing is certain: there must be a lot of particle physicists wishing they could swop their giant accelerators for a simple bubble chamber.

The other recent cosmic ray story is equally exciting, and equally entertaining. But this time the traditional roles of experiment and theory have not been swopped around. Indeed, the story is one of a triumph of ingenuity over adversity, and of imaginative theory over less than perfect experimental facilities.

Most of the best (or at least, most famous) cosmic ray detection centres are built high up on mountains, simply to get closer to the primary cosmic rays. But if you live in Leeds, mountains are fairly hard to come by. Nothing daunted, a team at Leeds University recently established one of the few cosmic ray detector arrays situated near sea level, and had a look to see what they could see. To almost everyone's surprise, what they could see was, in one vital respect, quite different from the view of cosmic rays provided by the supposedly better mountain-top experiments.

One of the ways of interpreting the information contained in cosmic rays is to plot out a 'spectrum' of the energies of the particles arriving at the Earth. When these energies are plotted out, various kinks are revealed in the graphs, indicating certain energies that are relatively favoured by whatever mysterious process it is that makes the cosmic rays. But the Leeds graphs, it turned out, contain an *extra* kink which has never been noticed in mountain top observations. And it is the explanation of this kink that puts the Leeds team into the history books.

There were three members of this historic team, Dr Walter Kellerman, Dr Gordon Brooke and Mr (now Dr) John Baruch. Delightfully, they have offered one explanation each to account for this new, sea-level kink. Dr Kellerman pointed out that the kink could be explained as the contribution of quarks to the overall spectrum. The quark is an interesting particle in its own right, having a charge of either $\frac{1}{3}$ or $\frac{2}{3}$ of the electronic charge,* and being, so the phy-

* Confusingly, some physicists have recently come up with a suggestion that quarks may even have 'whole number' charges. (This makes me happier than ever that I trained as a theoretical astronomer, not as a particle physicist!)

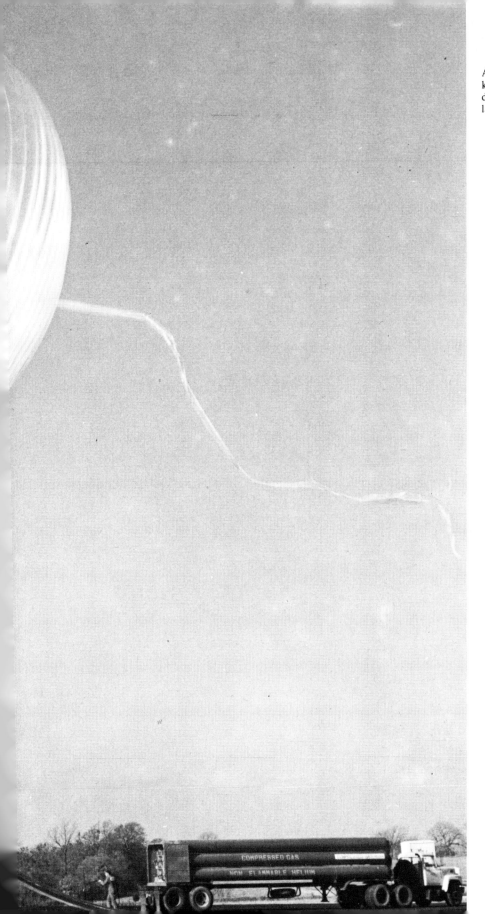

A helium-filled balloon of the kind used to carry astronomical detectors above the obscuring layers of the atmosphere

sicists believe, the fundamental particle from which other 'elementary' particles are built up. As the accelerators have got bigger and more expensive, the particle people have smashed elementary particles into more and more other particles, and always find that to explain what happens they need quarks; embarrassingly, however, even the biggest of their machines has not yet revealed a genuine quark to their puzzled gaze. So Dr Kellerman's interpretation of the Leeds kink would be the most pleasing to the theoreticians, if it turns out to be correct. But their hopes must be slightly dampened by Dr Kellerman's failure to convince his own colleagues that they have found the elusive quark.

Mr Baruch suggested at the time of the discovery (in 1973) that the kink could also be explained by the presence of another particle which theorists predict, the as-yet-hypothetical particle rejoicing in the name 'intermediate vector boson'. But Dr Brooke has said all along that he thinks there might be a 'conventional' explanation. More experiments at sea level (at Leeds and elsewhere) are continuing; again, the puzzle may well have been resolved by the time these words are in print. Whatever the outcome, though, there is no doubt that cosmic ray astronomy can be highly entertaining, and that the subject still has room for people with good imaginations and a fresh approach to science. Surely Lewis Carroll, at least, would have been delighted to know that at one of the frontiers of late twentieth-century science experimenters are busily engaged in the hunting of the quark, but that at present all that they can say for certain is that some quarks may be bosons!

Lewis Carroll might also have felt at home with the concept of tachyons mentioned earlier—it was one of his characters, the Red Queen, who said something to the effect that 'here, you have to run as fast as you can to stay in one place; to get anywhere else, you have to run much faster...'

This chapter has talked about some of the fantastic concepts of cosmic ray physics; but I have saved to the end perhaps the most fantastic concepts of all. If tachyons do exist, they are travelling backwards through time; and if they can provide a record of their existence by interacting with ordinary matter in our world, then at least in principle it seems that a 'signal' could be sent backwards in time, providing information about events which are in the future of the person who receives the signal. Isaac Asimov provided a picturesque view of the implications of such 'time travel' in a series of science fiction stories about a mysterious substance called Thiotimoline which dissolved just *before* water was added to it—the stories were written long before any hint that tachyons existed, but are well worth reading if you can find them.

All this is the sort of thing which enlivens many a discussion among astronomers during a coffee break, or over a beer or two, but is not usually taken too seriously. But in these days when strange phenomena (such as the claimed powers of Uri Geller) are taken seriously by some scientists, it is perhaps not too surprising to find at least one scientist looking seriously at the implications of the existence of tachyons for explanations of everyday events in the world about us.

Professor Jack Sarfatti, of the International Centre for Theoretical Physics in Trieste, Italy, is one of a handful of mathematicians now investigating the full implications of the concept of quantum mechanics; this depends on the postulate that, at a fundamental level, the observer of physical phenomena is an integral part of what he is observing. The concept has a long pedigree—much longer than the name 'quantum mechanics'—and edges into the realms of philosophy, where we start worrying about 'what happens to things when no one is looking at them? But we can begin to see why it is important if we take up the suggestion made by Professor J. H. Wheeler and replace the idea of 'observer' with that of 'participator'.

According to a mathematical formulation developed by Sarfatti, the 'participator' literally interacts with physical processes at a fundamental level: that is, the participator is affecting the behaviour of electrons and atoms, through a 'tachyonic interaction'. When a quantum particle might, within the laws of physics, behave in one of several alternative ways in response to a stimulus, its actual response will depend on 'the volition of the participator', as Sarfatti puts it. You may think this means that you could make electrons perform to your instructions by 'thinking at them'; but, as Sarfatti says: 'Generally the collective will of the participators is unfocussed and incoherent, giving the seemingly random character of quantum probability.'

But, you might well ask, what if some individual came along who did possess a high level of control over whatever mind process it is that has the tachyonic interaction with quantum effects? Sarfatti has the answer—you've all heard about Mr Geller already!

It may sound like science fiction—for all I know, it may even *be* science fiction—but don't forget that most of the current accepted astronomical theories about our surroundings must have sounded like science fiction or fantasy when they were first propounded. An alchemist of a bygone age might not have been too surprised by a theory that the Sun gains its energy from the transmutation of the elements, but the concept would certainly have seemed odd to a respectable astronomer of the mid-nineteenth century. That is, however, the best modern theory of what goes on inside the Sun—although there is, as we shall see, still some room for improvement in the details of the theory.

8 The Solar Neutrino Puzzle

In 1926, the great pioneering astrophysicist Eddington wrote: 'It is reasonable to hope that in a not too distant future we shall be competent to understand so simple a thing as a star'. At that time, the hope seemed perhaps a little optimistic. The nearest star, and the one we ought to know most about, is the Sun; but in the mid-1920s it was still a real puzzle how the Sun could exist in its hot fiery state at all. Simple calculations show that a 'Sun' made of solid coal would have burnt out after a time much shorter than the age of the Sun, which is known from the fossil and geological record. For a time, it had been thought that the heat generated by the gradual collapse of a cloud of gas into ever more dense conditions might provide the power source of stars—conversion of gravitational potential energy into heat, in much the same way that a bouncing ball warms up (slightly!) as its energy is dissipated as heat. But that energy source is not sufficient either, to account for the great age of the Sun (about four and a half thousand million years) and the stars.

But over the four decades following Eddington's expression of hope, everything seemed to become much clearer. The key to the new understanding of what makes the Sun and stars 'tick' was the development of an understanding of nuclear fusion, first in theory, then in the awesome practice of the hydrogen bomb—a miniature star.

This story is not so much a part of the new astronomy, and a full account of the processes of nuclear fusion in stars, and the synthesis of more complicated elements from hydrogen and helium in stars (stellar nucleosynthesis) can be found in many existing books. But the story is worth mentioning briefly for its own interest—and it also leads on to one of the greatest puzzles of the new astronomy, in which the complacency of the stellar theorists has recently been shattered by the discovery that the Sun may not, after all, 'tick' in quite the way their theories suggest it should.

Hydrogen is the simplest and most abundant element in the universe. The hydrogen atom contains one proton (positively charged) and one electron (negatively charged); helium (the next atom in complexity) has two protons and two electrons. Atoms also contain uncharged particles called neutrons, and when we are thinking about nuclear fusion it is the neutrons and protons which are important; they form the nucleus of each atom, whereas the electrons are tied more loosely, and further away from the centre of things. Indeed, in the conditions at the heart of the Sun, with temperatures upwards of 15 million degrees and pressure of a thousand million atmospheres), the electrons cannot be said to be tied to any particular atomic nucleus. Instead we have a sea of positively-charged nuclei and negatively-charged electrons, called a 'plasma'.

Considering just the nuclei, the steps up the fusion ladder are, at least for the first few rungs, quite simple. Two hydrogen nuclei (protons) can get together to form a nucleus of deuterium, or 'heavy hydrogen' (1 proton, 1 neutron) by emitting a unit of positive charge (a positron, or positive electron). Deuterium then combines with another hydrogen nucleus to produce helium-3 (2 protons, 1 neutron) and two nuclei of helium-3 can get together to produce one nucleus of helium-4 (2 protons, 2 neutrons) plus two hydrogen nuclei.

With two hydrogen nuclei going back into the 'pool', the net effect is that four protons have been converted into one helium nucleus, consisting of 2 protons and 2 neutrons. But the mass of a helium nucleus is *less* than the mass of four protons. Each proton has a mass of 1·008 atomic mass units (amu); but the helium nucleus has a mass of only 4·003 amu (*not* $4 \times 1·008 = 4·032$ amu). Somehow, 0·029 amu of matter have gone—almost one per cent of the original mass. Where have they gone? Einstein's famous equation $E = mc^2$ is so familiar as to be a cliché, but it holds the answer. For each helium nucleus created in this way, 0·029 amu have been converted into energy. And since m is the mass in Einstein's equation and c is the speed of light (3×10^{10} cm per second) the energy E quickly becomes quite large.

In round terms, the Sun is 'losing' 4·5 million tonnes of mass as energy every second, producing 4×10^{33} ergs of energy each second. Even this mass loss is no problem to the Sun, which has about 2×10^{33} grams to play with, and could burn fuel at that rate comfortably for a hundred thousand million years—ample time to explain the geological and fossil record. But it may not get the chance. As stars age, they must contract and become more dense at the centre, and then conditions can become right for more steps in the fusion ladder to be climbed. Elements more complicated than helium can be built up (including the carbon and oxygen so important to our bodies) and may eventually be dissipated across space when some stars explode, as novas or supernovas. This explosive end does not come to all stars; but, as we saw earlier, every atom of our bodies, and indeed of the whole world, has been through this process of stellar pressure-cooking. It is literally true, as the song 'Woodstock' says, that 'we are stardust... billion year old carbon'.

The only real evidence that these processes do go on, as described, in stellar interiors is that when you put all the calculations into a numerical model of a star the theory 'predicts' that a star with the mass and age of our Sun, for example, will be the same size as our Sun and have the observed brightness of the Sun. That is pretty convincing evidence; the equations involved are so complicated that it's very difficult to believe they could give the right answers by accident. Forty years after Eddington's statement of hope, the astrophysicists were sure they had the answers, and that only the details of stellar 'nucleosynthesis' remained to be filled in. A young re-

Overleaf left: The 158-inch Myall Reflector at Kitt Peak. Overleaf right: The Orion Nebula

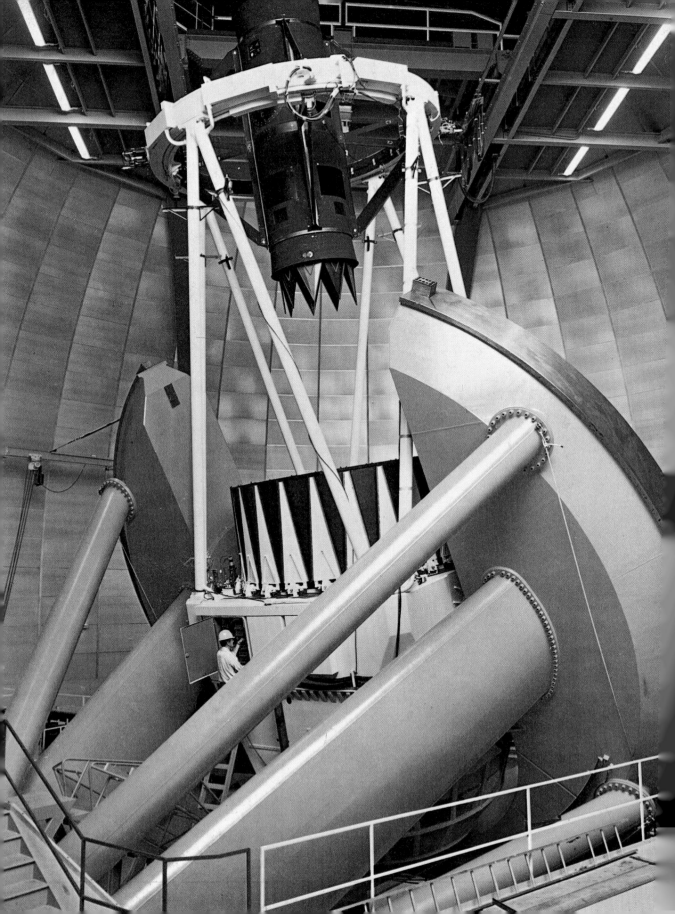

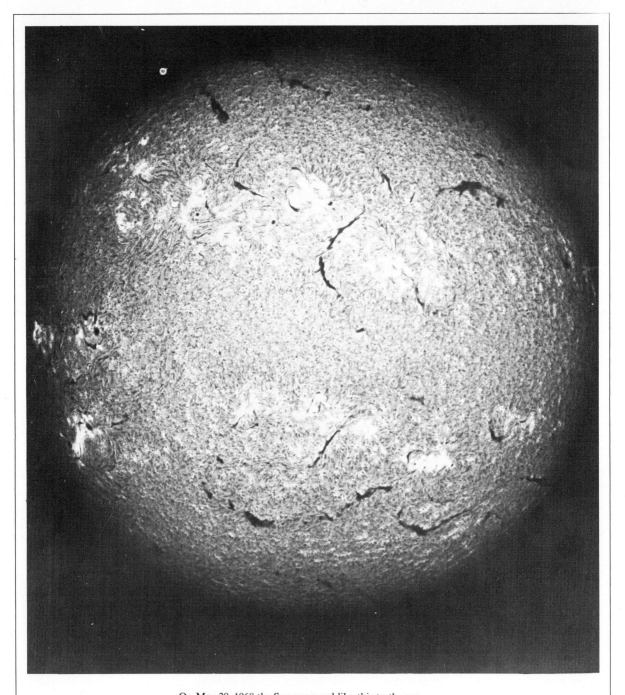

On May 29, 1968 the Sun appeared like this to the eye
of a new telescope recently installed at The Pennsylvania State
University. Taken during a relatively quiet period, the white
swirls or 'plages' indicate areas where flares that often
disrupt shortwave radio transmission are likely to appear. The
narrow black lines are other solar outbursts known as prominences.
The picture was taken with a filter which filters out all light
produced by the Sun, except for a very narrow band in the visible
spectrum known as the hydrogen alpha line

searcher starting out in astronomy in 1966 would hardly have turned to the study of stellar interiors as an exciting field to do research in; but he would have been wrong.

The reason for this is that it has now become possible to make a more-or-less direct measurement of something happening at the centre of the Sun. At least, according to all the standard theories of stellar nucleosynthesis something measurable *should* be reaching the Earth—but there is no evidence that it has yet been found. And that has really upset the stellar applecart.

The concept of a particle that could travel from the middle of the Sun out through the Sun and space to Earth without being affected by anything on the way would probably have astonished even Eddington. But these elusive particles, dubbed neutrinos, are an essential feature of the best theories of nuclear physics. According to these theories, many nuclear reactions just cannot work without neutrinos around—it is because the reactions *do* take place that the physicists have had to invoke the existence of neutrinos to explain their results.

But the neutrino is certainly an odd bird. It has no mass, travels at the speed of light and carries 'spin'. The peculiarities need not concern us here; all that matters is that neutrinos are so loth to interact with other particles (such as neutrons and protons) that only one neutrino, out of every 10^{11} produced by nuclear reactions inside the Sun, fails to escape completely into space. To a neutrino, indeed, the fiery Sun, and the solid Earth are hardly any more of an obstruction than empty space itself.

Elusive they may be, but there is no denying the importance of neutrinos. According to the theorists, three per cent of the energy radiated by the Sun is in the form of neutrinos, and just about 10^{11} neutrinos from the Sun cross every square centimetre of the Earth (including my body and yours!) every second. If we could catch some of them, their presence would provide direct confirmation that the nuclear reactions going on in the Sun are ticking over just as the theories 'predict'.

But there's the snag—how do you catch an object that scarcely deigns to notice the existence of the Sun, and would require a layer of lead several light years thick to stop it? The answer is that you cannot catch *one* neutrino—but with 10^{11} crossing each square centimetre every second, there is a good chance that a few of them might interact with more solid particles, given the opportunity.

Providing the opportunity is almost a story in itself. The standard neutrino detector is best sited deep below ground (usually in a disused mine shaft) so that the rock will keep out everything except neutrinos. Even then, you need as big a detector as possible, and a tank of suitable fluid holding about 100,000 gallons was used in the first attempts to trap solar neutrinos. What do you put in the neutrino trap? As it happens, nothing too sophisticated is needed. Neutrinos ought to be fairly partial to ordinary cleaning fluid (carbon tetrachloride); about one quarter of the natural chlorine atoms will be the

Overleaf: The Crab Nebula

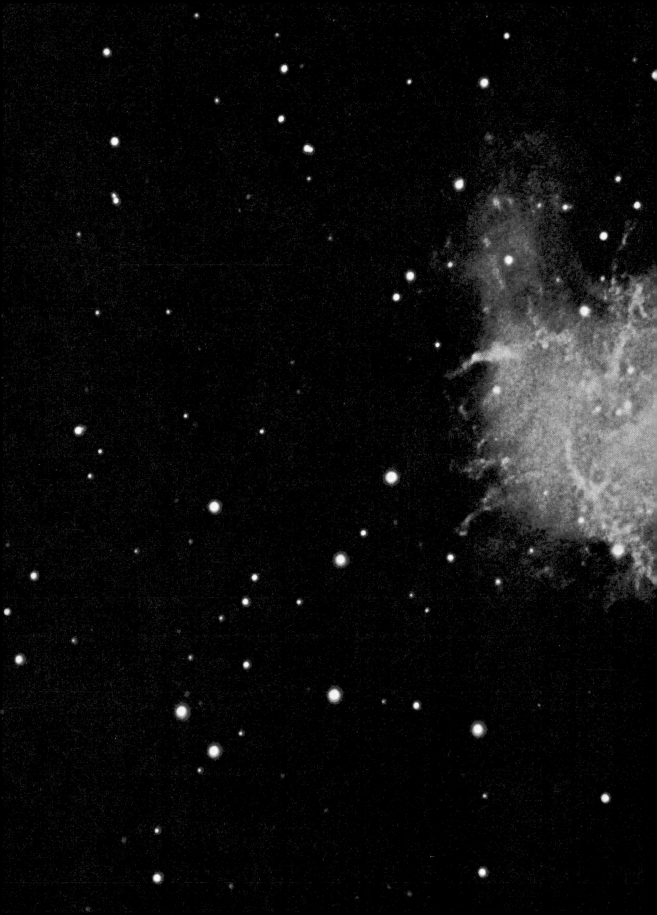

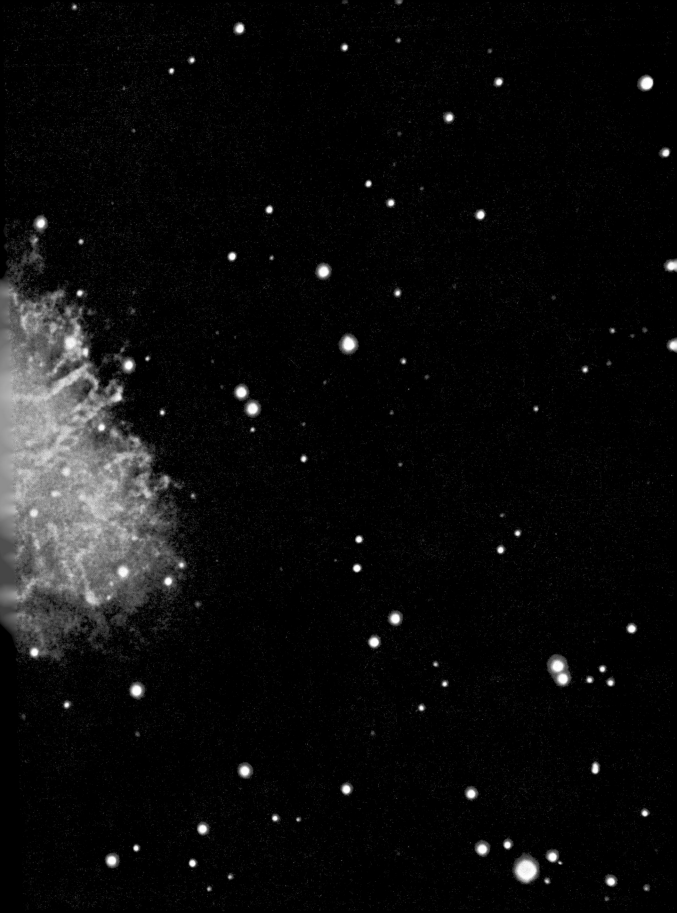

isotope chlorine-37 (the rest are chlorine-35, containing two less neutrons but the same number of protons) and a neutrino can interact with chlorine-37, to produce argon-37 and an electron. So fill your tank up with cleaning fluid, wait for a month or two, and count to see how many atoms of argon-37 you have in the tank. This will be the same as the number of neutrinos that have stopped in the tank.

Easier said than done, but it is possible as long as there are a dozen or so argon-37 atoms to work with. Indeed, when baths of this fluid are placed near nuclear reactors, argon-37 is produced just as predicted—neutrinos do exist. But what of the solar neutrinos and the neutrino traps deep underground? Here the story develops into a real puzzle. The complexity of the problem is indicated by the units with which the experimenters try to measure their catch. One solar neutrino unit (or SNU) is equivalent to 10^{-36} neutrino captures per second per chlorine-37 atom. Put another way, if the neutrino trap contains 10^{36} chlorine-37 atoms, and out of the whole tank one neutrino is captured every second, then the capture rate is 1 SNU. The smaller the amount of chlorine-37, the longer the wait before one neutrino is likely to be captured.

What do the theories of the workings of the Sun predict? About 1964, the first proposals were that about 28 SNU should be detectable at the Earth, and this fairly high figure (by the standards of neutrino hunters) encouraged proposals for experiments to be put forward. Even before the experiments got under way, however, the theorists revised their estimates downwards; in recent years they seem to have settled around 6 to 8 SNU. But that still ought to be enough.

The first experimental results, in 1968, reported no more than 3 SNU; and since then improved experimental evidence places a limit of 1 SNU on the amount of neutrinos reaching us from the Sun. Indeed, it seems quite likely from this evidence that there are no solar neutrinos reaching us at all.

Where does that leave the theorists? In something of a turmoil, because if there are no solar neutrinos (or even if there are some but less than predicted) then their theories of the nuclear physics of stellar interiors must be wrong. Taking one key example, the neutrino experiments suggest that the centre of the Sun may be at least ten per cent cooler than was thought—and that is a big difference for a theory which had seemed only four years earlier to have produced all the answers.

Assuming the observations are correct, there are only two possible answers. Either the theories of nuclear physics are wrong (and that is a very worrying prospect to all of us today, when the prospect of developing fusion power in reactors on Earth is one of the main hopes of beating the energy crisis), or the nuclear fusion reactions just are not going on in the Sun at present. That seems ridiculous at first sight—after all, the Sun is still there, still as hot as ever. But that answer looks the best way out of the dilemma, and it is not so crazy as it seems at first sight.

Right: The Brookhaven National Laboratory solar neutrino detector at the Homestake Goldmine in Lead, South Dakota. The tank contains 100,000 gallons of carbon tetrachloride

Overleaf: Top left—The planetary nebula NGC6543. Bottom left—The Ring Nebula in Lyra, also known as M57. Top right—The Dumbbell Nebula, M27. Bottom right—The active galaxy M82

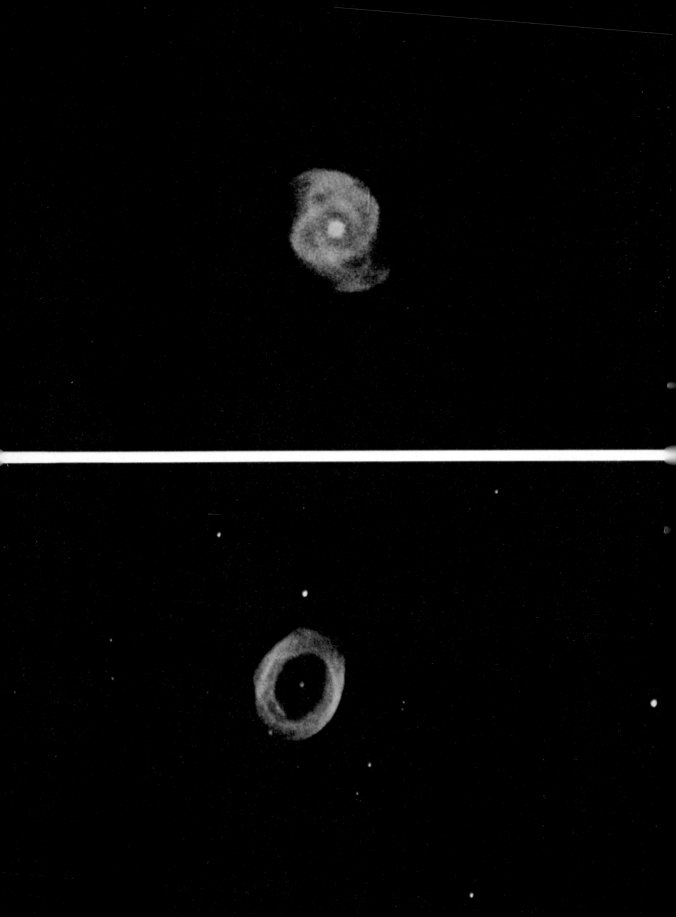

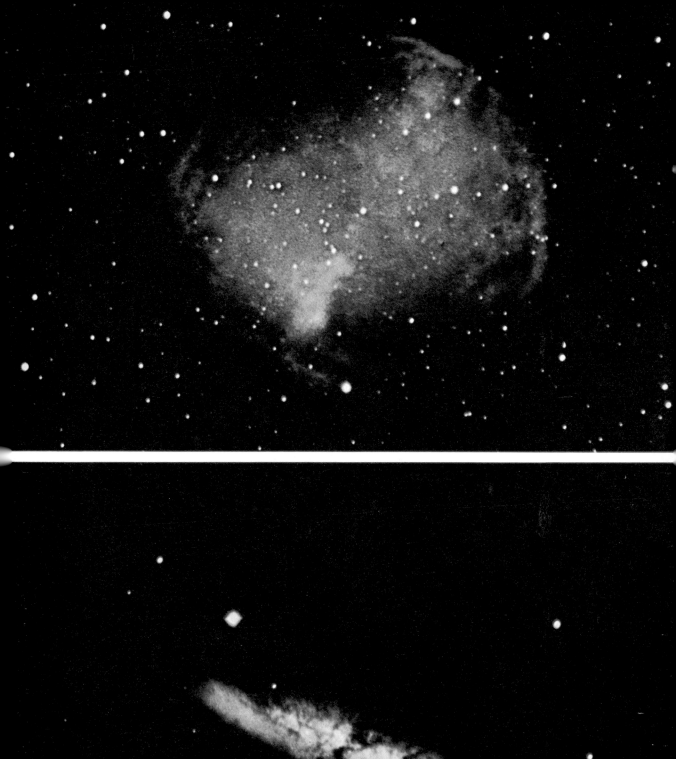

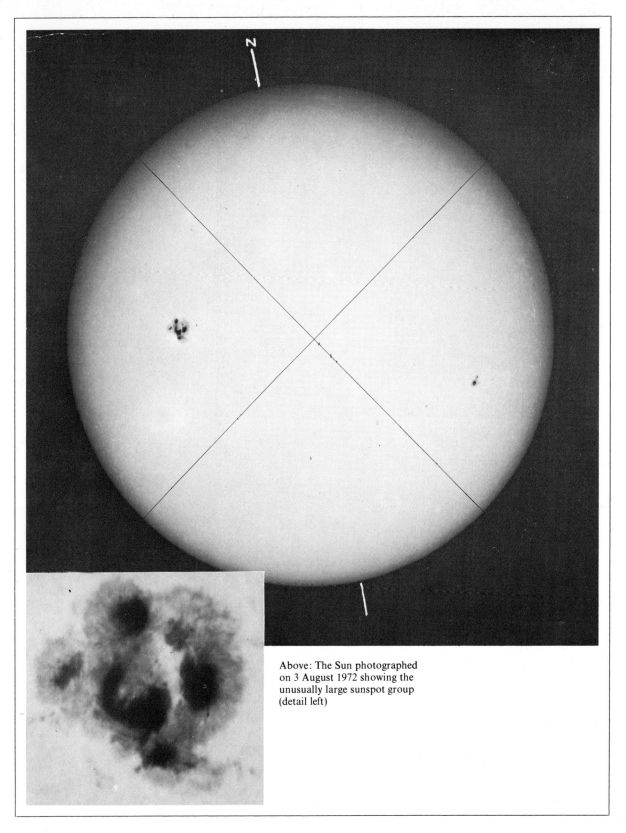

Above: The Sun photographed
on 3 August 1972 showing the
unusually large sunspot group
(detail left)

The point is that the Sun takes a long time to adjust to changes going on inside it. Switch off the nuclear fusion and the Sun does not just go out like a light; it settles down slowly, contracts a bit and gains some energy from its gravitational potential, and generally seems to mull over the situation for thousands—even millions—of years. In effect, the heat and light we receive from the surface of the Sun are an average, smoothed-out effect of solar conditions over the past ten million years. But neutrinos, travelling at the speed of light from the centre of the Sun, reach us after only eight minutes.

So that is what many people see as the answer to the solar neutrino puzzle. The warmth and heat we receive from the Sun tell us that nuclear burning has been going on for most of the past ten million years; the lack of solar neutrinos tells us that nuclear burning has not been going on since 1968, and may not have been going on for a couple of million years before that. Even if the nuclear burning does not start again for another couple of million years, humanity has nothing to worry about. Or have we?

Why should the Sun have gone off the boil, as it were? And how is this likely to affect us? The theorists have been quick to produce models purporting to explain the phenomenon, and some have shown that the Sun can go off the boil because of slight changes in the convection processes which carry heat outward from its centre. The result would be that the Sun cooled for about a million years, then contracted a little and warmed up again, with the nuclear burning restarting. Now, over the past two million years the Earth has suffered repeated ice ages—far more than normal, judging from the geological record of the past twenty million years. Is it possible that this phase of cool climate is related to a slight cooling of the Sun? The climatologists, in fact, have an embarrassment of explanations for ice ages, and are not entirely happy to welcome yet another suggestion*. But it is just possible, and fun to speculate about.

But perhaps fun is the wrong word. Could it be that the cold which killed off the dinosaurs is a result of changes in the Sun's heat? If so, it is certainly fun to speculate that the beginning of the rise of man himself might have been due to a tiny and temporary fluctuation in the workings of the Sun. The fun is diminished, however, by the other side of the coin. Suppose that the nuclear burning is now switching back on and the Sun will warm up again—or that the cooling is going to continue for another million years, with more and worse ice ages. Either way, it looks horribly as if a tiny fluctuation in the workings of the Sun could be the death of our civilisation, as well as its godfather. Such sobering thoughts left aside, however, astrophysicists are still left with a lesser sobering thought: fifty years after Eddington's statement, we still do not, after all, 'understand so simple a thing as a star'. It's worth keeping that in mind, perhaps, when we look at some of the more bizarre phenomena which have come to light through the explosive growth of the new astronomy.

Overleaf: A long exposure photograph of the Orion Nebula (compare with the shorter exposure photograph on page 67)

* This is discussed in my book *Forecasts, Famines and Freezes* (see Bibliography).

77

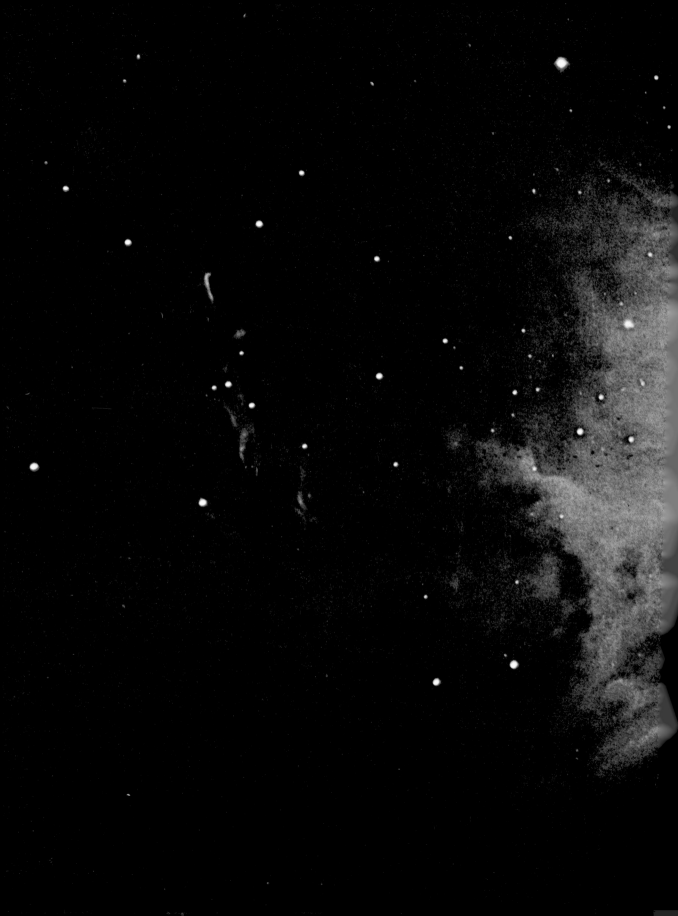

9 Sco X-1, The Enigmatic X-Ray Source

Whatever their nature, be they quarks, tachyons or other particles, the presence of large numbers of cosmic rays in our galaxy shows that it is an active place, where violent events are not uncommon. Indeed, this is perhaps the most important thing about the new astronomy: almost without exception, it is the astronomy of high-energy astrophysics, involving the study of explosive events and the debris of exploded stars, or even exploded galaxies.

Within our own galaxy, the most energetic of the more-or-less steady astronomical sources of energy are X-ray stars—stars so energetic that most of their emitted radiation is in the form of X-rays. The first of these sources was discovered by accident (another example of astronomical serendipity) only just over ten years ago. This source, known as Sco X-1, is widely regarded as the archetype of the class of X-ray stars. But it is rather a curious archetype, since even after ten years of study it remains incompletely understood, although more recently discovered X-ray stars can be explained quite satisfactorily. Shrouded though it is in an aura of enigmatic mystery, Sco X-1 remains just about the most intensively studied X-ray star. The story of its study shows the development of the new science of X-ray astronomy in much the same way that the study of Cygnus A (Chapter 2) closely followed the development of radio astronomy. In that sense, then, Sco X-1 is certainly the archetype, and worthy of a closer look than most X-ray sources.

When scientific historians come to write the history of the 1960s, they will probably accord as much scientific significance to a 'Moon rocket' launched on June 18, 1962, as to the epic voyage of Apollo 11. Paradoxically, that 'Moon rocket' was, strictly speaking, a failure. The prime objective of the mission was to carry X-ray detectors above the atmosphere, point them at the Moon and, hopefully, detect X-rays produced when solar cosmic rays struck the lunar surface. No evidence was found for such emission, and none has been found since. But as the detector scanned across the sky during its brief flight, it recorded intense X-ray emission (at wavelengths of 2 to 8 Ångströms), far stronger than that from our own Sun, coming from a point source in the constellation Scorpius. In addition, a faint background of X-rays was detected from all directions scanned.

The theoreticians were caught on the hop by this discovery—none of them had predicted the existence of long-lived point sources of X-rays stronger than the Sun, and yet here was such a source—an X-ray star emitting 10^{16} times as much X-ray energy as the Sun (and 100,000 times as much energy as the *total* from the Sun at all frequencies!). Needless to say, the discovery encouraged a search for more X-ray stars, and several were found, despite the problem posed by the need to fly instruments above the atmosphere, in rockets and balloons, with no indication of where they should be pointed to get the best results. As far as Sco X-1 was concerned, the next major advance (really the next major advance in the study of discrete sources of X-rays) came in 1966 when, thanks to a joint programme of study by Japanese and American astronomers, Sco X-1 was identified with a star visible in ordinary light—at optical frequencies.

This helped in two ways. First, astronomers on the ground could at last monitor what was happening to the star; and second, as with any other astronomical object, measurements at many different frequencies help to build up an overall spectrum, which can reveal details of what is going on in the source. Hardly surprisingly, the star now revealed as Sco X-1 is far from normal. To start with, it is very blue (in astronomers' jargon, its spectrum shows a large ultra-violet excess); it also shows a very rapid 'flickering' variability, super-imposed on longer term flaring in which it can brighten dramatically in a short time before calming down again.

There is really only one way in which this kind of behaviour can be explained, together with the production of such enormous amounts of energy in the star. This explanation involves conversion of gravitational potential energy into heat, or kinetic energy; it also requires that there must be *two* stars associated with the X-ray source—that it is a binary. In this situation, gas can be pulled off from the larger star of the pair by tidal effects, and spiral down onto the smaller star. As the gas falls in the gravity field of the small star, it speeds up. In other words, it gains kinetic energy at the cost of gravitational energy. Provided the small star is compact enough, a great deal of energy can be liberated in this way. So we have a situation in which the small star is surrounded by very hot gas, or plasma, which radiates at X-ray frequencies and is constantly being renewed by gas from the big star.

Such a system can keep going for a good while by human stan-dards, although of course there comes a time when the larger star has shrunk too far for the mass exchange to continue. So X-ray stars ought to be fairly short-lived in astronomical terms. But it is one thing to wave your hands about and explain in a vague way how to make an X-ray star; it is quite another to develop a proper 'model' of the source. How big are the two stars? How much gas needs to be transferred? How did they get to such a state in the first place? These and other questions nagged at theoretical astronomers through the late 1960s—and some of them have yet to be answered to everyone's satisfaction.

Before looking at one detailed model in particular, it is worth mentioning just how much of the whole electromagnetic spectrum is covered by the radiation from Sco X-1. The high-energy or 'hard' X-rays form one extreme, at very high frequencies. The shape of the X-ray spectrum looks very much as it should if the radiation were

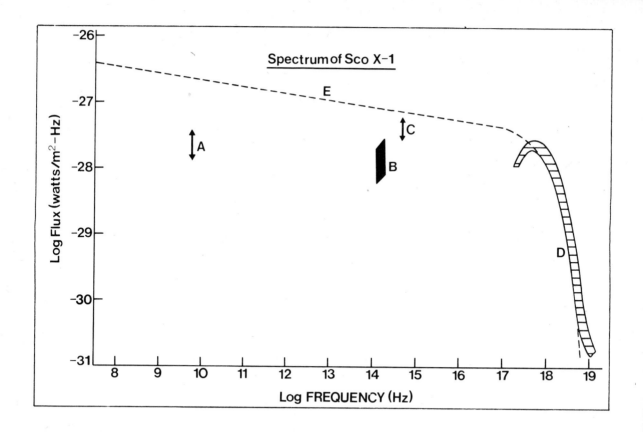

Spectrum of Sco X-1

coming from a hot plasma, at a temperature between 10 million and 100 million degrees absolute (10^7–10^8 K). Working down to lower energies (longer wavelengths or lower frequencies), the known spectrum includes the optical (visible light) emission; this has absorption lines superimposed which show the presence of helium, carbon, oxygen and many other elements, in a cooler region (about 10,000 K) further out from the hottest, most energetic part of the source. Sco X-1 has also been detected emitting at infrared frequencies; and at lower frequencies still it is known as a weak, variable emitter of radio waves. All this radiation provides a lot of information; and it is a fair indication of the difficulty of finding a good 'model' for Sco X-1 that all this information still partly defies explanation. But one interesting insight is provided by the radio emission.

In 1971 it was reported that the radio source can be resolved into a triple structure, superficially not unlike the kind of structure in Cygnus A and other radio galaxies, although on a very much smaller scale. One of these three components lies exactly on the blue star identified with Sco X-1, and this component varies over periods of a few hours. The other two components lie on either side of the star, but close by it. This seems to hint very strongly that some kind of explosion occurred in the source a fairly short time ago (by astronomical standards). Something like a nova, perhaps? That would agree very well with what we know about energetic sources like the

Above: The spectrum of Sco X-1. A—radio measurements; B—infrared observations; C—optical; D—X-ray emission; the dashed curve E is the spectrum expected if the electromagnetic radiation is coming from a plasma at a temperature at a few tens of millions of degrees. Right: By interpreting the periodic variations in the flickering of Sco X-1 as vibrations of a hot plasma, we find the minimum gravity of the underlying star. Only objects with gravity in the range above the dashed line can explain the observations—that is, white dwarf or neutron stars

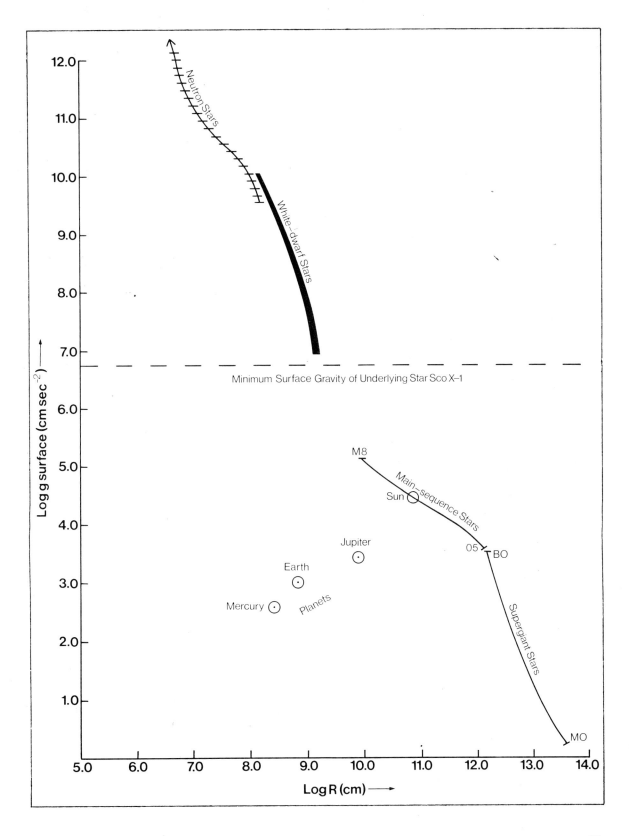

Crab Nebula. Better still, some of the latest radio observations show that the radio emission from Sco X-1 is rather like the radio noise from a pulsar, but without any pulses. That is not quite as nonsensical as it sounds. Remember, the best explanation for pulsars is that they beam out pulses in the equatorial plane of a spinning neutron star, so the pulses would not be detectable if the Solar System and the Earth did not lie in the path of this beam. But the lower-level, general radio 'noise' of pulsars could well be broadcast in all directions, independent of the beaming process.

Indeed, some astronomers confessed to a feeling of relief when it was discovered that Sco X-1 might well be a kind of pulsar, beaming its pulses away from the Earth. The point is that most pulsars should, statistically, *not* be beaming their pulses directly at the Earth, and it had seemed an almost improbably great stroke of fortune to find that the Crab pulsar, the youngest and most energetic of all the pulsars, was aligned just right for observation from Earth. Finding other energetic sources beaming energy in other directions makes the whole situation look far happier; perhaps the laws of probability are not being stretched outrageously after all.

But that is really an aside; we cannot yet say for certain that Sco X-1 is associated with some kind of pulsar. However, the evidence is certainly suggestive. A neutron star, or perhaps a white dwarf, is just the kind of compact star needed for the gravitational energy conversion to work satisfactorily. And studies of the 'flickering' of the light from Sco X-1 have made it possible to narrow down the remaining area of doubt about the detailed nature of the object quite considerably.

One of the snags about being a theoretical astronomer is that you have to wait for someone else to produce observations which you can try to interpret in new ways. But on the other hand, the observers are usually pretty generous with their data. In 1969, when I was working with a group of astronomers in Cambridge, we were given a copy of the information obtained by astronomers at the Mount Wilson and Palomar Observatories during five nights of continuous monitoring of Sco X-1, using variously the 60-inch, 100-inch and 200-inch reflecting telescopes. These data came in the form of a deck of punched cards, with the brightness of the star recorded as numbers at intervals of 5 seconds (or 15 seconds for one night's observations). This sort of data is just right for analysis in a computer, and several people (including the original observers) had searched to see if they could find traces of periodic variations in the light from Sco X-1. For, as we saw in Chapter 4, periodic variations tell us a lot about the nature of astronomical objects.

But all those investigations had one thing in common; they involved using the longest stretches of observations possible because if a small periodic variation *is* present, it is best seen by adding up (in the proper statistical fashion) a long 'run' of observations. But when we came to investigate the data from California, I encountered the good fortune which often seems to smile on astronomers.

The computer program with which we were working in Cambridge at that time could find periodic variations even if they were hidden by random effects ('noise') from the unaided eye, using a technique called 'power spectrum analysis'. But our program just could not handle a whole night's worth of data in one gulp. So we were forced to study shorter stretches of the California observations—which resulted in the remarkable discovery that some of those short stretches showed strong periodic variations, while others were, from our point of view, completely blank.

This seemed a very curious state of affairs: how could it be that sometimes the flickering from Sco X-1 has a periodic structure, while at other times it is random? A clue to the answer came when I discovered that the strongest periodic variation occurred just after one of the strong outbursts, or flares, from the star. It looked as if the flare had in some way set off the periodic variation (which repeated over a three-minute period) almost like the ringing of a bell that had been struck by a giant hammer. This gave us a clue to the structure of the Sco X-1 system. If it really is a binary system, we argued, then the flare could correspond to an outburst of the larger star dumping an unusual amount of gas onto the small star. That would stir up the hot plasma cloud around the small star, and make it oscillate at its natural frequency rather like the ripples spreading from a stone tossed in a pool.

Now, the oscillation of such a hot plasma will depend both on the physics of the plasma, such as its temperature, and on the strength of the gravity field of the star round which the plasma circles. Studies of the spectrum of Sco X-1 tell us what the plasma is like, so, using the three-minute period as the oscillation period, we could work out the minimum gravity needed to explain the observations. The answer was not unexpected—it turns out that the small star must be a white dwarf or a neutron star, just as the theories had already predicted in a general sense. But this was the first direct confirmation that the Sco X-1 system contained such a very dense, compact star.

In the past four years X-ray astronomy has developed dramatically, and several sources are now known to be associated with such compact stars, as we shall see in the next chapter. The study of Sco X-1 itself has also made progress. One of the best lines of attack has been the development of theories to explain in more detail what happens to the gas as it falls from the large star to the smaller; it cannot fall straight down, but must spiral inwards as it loses angular momentum. That could explain another effect apparent in the California data I studied, namely that the three-minute variations themselves could be seen most clearly in stretches of data about half an hour apart; it looks rather as if the 'vibrating' gas which we see in the flickering of Sco X-1 is orbiting in a ring around the underlying star, so that it is periodically eclipsed.

Once again, the numbers fit in with the white dwarf or neutron star explanation. Recently other observers have reported detecting

similar periodic variations in the light from Sco X-1. Piece by piece, a picture has been put together which looks remarkably similar to the picture astronomers have of the so-called 'dwarf novae'. These systems, which have been studied for many years, show weak pulses of periodic variations, which sometimes disappear and which change in period according to the brightness of the star. The stars of these sources are in binary systems, and they get their name because they are subject to explosive outbursts which are not quite dramatic enough to be graced with the name novae. One attractive theory is that the outbursts are related to the sudden dumping of matter onto the dwarf star from its companion, in much the way as I have suggested to account for the behaviour of Sco X-1. The outbursts occur at irregular intervals, but about once a month; again rather like the flaring of Sco X-1.

So astronomers are beginning to penetrate the mystery surrounding Sco X-1, the archetypal X-ray source. Before they could find out about X-ray sources in general, though, they had to await a new technical development—a satellite devoted to X-ray astronomy, which

Early days at the Institute of Theoretical Astronomy, Cambridge—the entire staff photographed in the summer of 1967, the year IOTA was started. Back row: Mary Gribbin (then IOTA's assistant librarian) and the author (then a computer operator), with other computer staff. At the centre of the middle row, Professor (now Sir) Fred Hoyle, then Director of IOTA, sits next to Professor Margaret Burbidge (sleeping). Luminaries of the front rank include Professor Willie Fowler (with tennis racquet) and, to his right, Professor Geoff Burbidge; at the extreme right of the picture in the front row is Dr John Faulkner, former cycling champion

could scan the skies continuously. No longer do they have to depend on brief rocket flights to investigate the universe at X-ray frequencies, and this has led to a veritable explosion of X-ray astronomy. But, paradoxically, the observations which have done most to provide an insight into the place Sco X-1 occupies in the X-ray star family, have not been made at X-ray frequencies at all. They were observations of an X-ray star, true; but observations made at radio frequencies. To find out why these observations were so important, we must first look at the overall development of X-ray astronomy since the launching of the first X-ray satellite.

10 X-Ray Astronomy's Big Step Forward

Up until December 1970, the main obstacle to the development of X-ray astronomy was the Earth's own atmosphere: it was possible to get quick glimpses of X-ray sources in space from rockets and balloons, and in the case of Sco X-1 the identification of the source with a visible star meant that it could be studied optically from the ground. But apart from that, there was nothing. And this sporadic peeping at the X-ray sky hardly qualified as a branch of astronomy in its own right. With a mere handful of sources known, and no way of searching adequately for other sources, there were huge and literally unimaginable gaps. The big step forward in the science of X-ray astronomy came with the launching of an American satellite from an Italian launching pad just off the coast of East Africa, at the end of 1970—the first satellite dedicated wholly to X-ray astronomy.

This launching site was chosen because it is near the equator, where the spin of the Earth gives the biggest kick to a rocket launched on a path from west to east. In honour of the site, the satellite, Explorer 42, was given the name 'Uhuru'—the Swahili word for freedom. This has proved most appropriate; for X-ray astronomers are now freed for good from the obscuring blanket of the Earth's atmosphere.

Uhuru was designed to produce a sky survey, the satellite spinning on its own axis as it orbited the Earth, so that within a few months its scanners had swept across the whole sky. This meant that it could detect every strong X-ray source, the snag being that the spin could not be stopped, so that the scanners could only 'look' at a source very briefly as they swept around. But first things must come first, and there is no point in having accurately pointing scanners if you do not know where to point them.

Within weeks, Uhuru was producing observations that stirred the astronomical community into feverish activity. The first of the surprises came when Cygnus X-1, a source already known from rocket and balloon observations, turned out to be behaving like an X-ray pulsar, apparently flashing at 15 pulses per second. In the same batch of early results, Uhuru had found several previously unknown X-ray sources outside our galaxy, and by early 1971 astronomers were beginning to regard the observations from the satellite as solid, useful new advances. But then the discoveries began to get more exotic, and the theoreticians were thrown into some confusion trying to explain them all.

The discovery that other galaxies and clusters of galaxies could be identified with X-ray sources was in some ways the easiest for astronomers to explain. After all, if objects like the Crab Nebula, Sco X-1, Cygnus X-1 and ordinary stars like the Sun (in its own quiet way) could produce X-rays, then there could well be a respectable average X-ray emission from a whole galaxy, observed from outside. This kind of explanation immediately puts the ball back in the court of our own galaxy: certainly there are enough bits and pieces of X-radiation produced within our own galaxy to make it noticeable as a whole at X-ray frequencies—but the interesting question is, what makes the bits and pieces so active?

Spurred on by the discovery of pulsations from Cygnus X-1, the Uhuru observers turned their attention to other sources in a search for similar effects. Sure enough, they found that another source, Centaurus X-3, also seemed to be 'pulsating', this time at intervals of about 5 seconds. But neither Cyg X-1 nor Cen X-3 showed any pulsing at radio or optical frequencies, and this came as a great puzzle at a time when pulsars themselves, detected by their radio signals, were just beginning to be understood. Was an X-ray 'pulsar' some new kind of object? Or was it a variant form of pulsar for which the developing theories had made no provision? The only sure thing seemed to be that, like Sco X-1, these sources must be associated with very dense, old stars. Just as the development of optical astronomy, earlier in the twentieth century, had allowed astronomers to build up an explanation of how stars are born and age, so X-ray astronomy had begun to open the way to an understanding of the old age and death of stars.

Throughout 1971 more observations of an ever-increasing number of sources came flooding in from Uhuru; and once Uhuru had pinpointed new sources, rocket flights could be used to study them further with different instruments. By the beginning of 1972, the picture was beginning to come into focus. There were, it seemed, two kinds of variable X-ray source. First, there were pulsars like the Crab pulsar, which produce very regular pulses through some rotation process involving the whole star. Secondly, there were the more erratic variables, like Sco X-1 and Cyg X-1; these might be explained as stars whose 'atmosphere' and surroundings produce changeable or short-lived periodic effects. One suggested explanation for the pulsations of Cen X-3 revives the pulsating white dwarf models which failed to explain radio pulsars; that explanation is still an acceptable one, although probably an oversimplification in its least elaborate form.

But the real key to understanding the new class of X-ray variables lay in the accumulating evidence that they were indeed binary systems—just as has been postulated (but never absolutely proved) for Sco X-1.* With a whole year's observations available, it became

* What seems to be conclusive evidence that Sco X-1 is a binary was published in 1975; this now opens the way to a full understanding of the archetypal X-ray source as a member of the same binary family as many other X-ray sources.

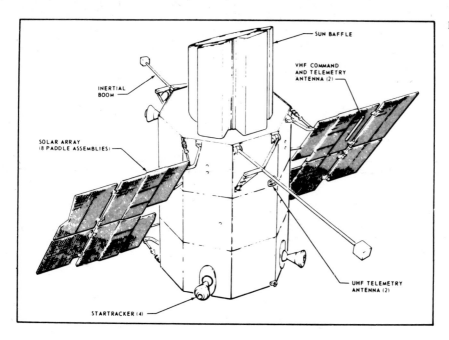

clear that Cen X-3, for example, was showing both long term variation, with a period of two days, and a rapid 5-second variation. This could only be a result of the orbital movement of the X-ray source around another star. What was needed next was accurate pinpointing of these sources; if any of them could be identified with optical stars visible through ordinary telescopes, the models could be tested. And this development was not long coming.

In the Autumn of 1971, two British groups applied the lunar occultation technique, which had proved so successful in pinning down radio sources, to the X-ray source GX3+1 (GX stands for Galactic X-ray source; the numbers are a shorthand form of its position on the sky). The technique required unprecedented accuracy of timing: two rockets had to be launched at just the right time so that they would have reached coasting flight, stabilised, and had time for their sensors to 'lock on' to GX3+1, before the occultation occurred. Even then, the work of the experimenters was just beginning: they had to calculate the exact position of the rocket in space at the instant of occultation, before they could pinpoint the arc on the sky corresponding to the edge of the Moon as 'seen' by the rocket. Some idea of the urgency and excitement generated by X-ray astronomy at that time can be gained from the speed with which these complex calculations were made. For the first of the two flights (from Australia), Professor Ken Pounds, of Leicester University, was presenting the new figures to a meeting in Bristol on October 1— just four days after the occultation flight itself! Appropriately, that meeting was being held to mark ten years of X-ray astronomy, since the discovery of Sco X-1.

Together with observations from a second occultation experiment

Right: How Copernicus pins down the position of an X-ray star. Top: The X-ray sensors track across the suspected position of the source in a zig-zag path. Middle: The most accurate Copernicus detector is used to search around the most probable site. Bottom: The final Copernicus position compared with the approximate position found by Uhuru

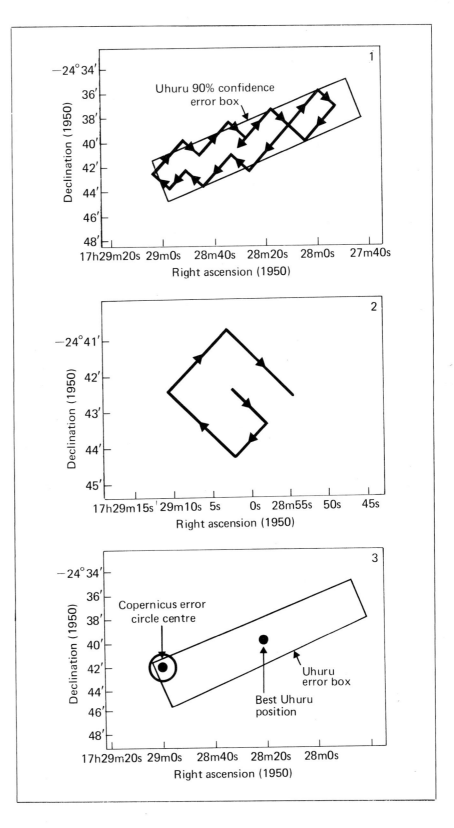

on October 24, these measurements located GX3 + 1 to an accuracy a thousand times better than the best previous measurements. But within another year, X-ray astronomers had a new satellite with X-ray telescopes in orbit, and were able to do better still.

This new satellite was Copernicus, so called to mark the 500th anniversary of that great astronomer. The principal experiment carried by Copernicus is an ultraviolet telescope, run by Princeton University; but it also carries a battery of X-ray telescopes operated by the Mullard Space Science Laboratory (MSSL), London. This secondary experiment rather stole the thunder from the main experiment, in the first months after Copernicus was launched.

The set of X-ray telescopes looks rather like a Gatling gun, and the different 'barrels' are switched in to provide different kinds of X-ray imaging, ranging down to pinpoint focus—which can, of course, only be used once the position of the source being looked at is known very accurately. The system was planned to be pointed with an accuracy of 0·1 arc sec, which is equivalent to pinpointing an object the size of your head from a distance of 400 miles. In fact it is three times better than planned, and points accurately to within 0·03 arc sec.

Uhuru and Copernicus really worked as a team, Uhuru spinning round and round discovering new sources, and Copernicus following up, looking long and hard at them. Thirty stars could be studied each year, in detail—an almost unbelievable situation for X-ray astronomers, who in the 1960s could only expect rather vague information about maybe five or six sources a year. But there is still plenty of scope for more satellites like Copernicus to study the hundreds of interesting objects still to be examined.

The Uhuru satellite has now come to the end of its life as an X-ray observatory. But by the beginning of 1975 a British X-ray astronomy satellite, Ariel 5, had been put into orbit (by an American rocket). First results indicate that Ariel 5 is working perfectly, and no doubt we will now see a new partnership, between Copernicus and Ariel, taking up where the Uhuru-Copernicus partnership left off.

The diagram shows just how Copernicus pins down the position of an X-ray source before beginning intensive studies. First, the Copernicus telescopes scan the entire 'error box' of the rough Uhuru position, in a series of zigzag steps. This shows which part of the box the X-rays are coming from. Then, the telescope with the best pointing accuracy is switched in and used to search in a spiral around the most active part of the Uhuru box. The final position shows the location of the X-ray source to within two arc min at the most, making it much easier to identify the source with an optical star.

Copernicus has also been used to test out the feasibility of lunar occultation observations from a satellite, and passed the test with flying colours. GX5-1 was located to within a 'box' only 6 arc sec by 20 arc sec; the most accurate previous position was 1·5 arc min across.

There is still another way of pinpointing an X-ray source. If it can be identified with a radio source, then the position of the radio source can be calculated, usually with greater accuracy than the X-ray measurements can give. This too narrows down the 'error box' and improves the chance of identifying the source with an optical star.

In some cases, all the pieces fit together beautifully. Hercules X-1, for instance, has been identified with a double star system, and studies of the motions of the two stars make it possible to work out their masses: the X-ray 'pulsar' must be about half as massive as the Sun, just right for a neutron star. But one system, Cyg X-1, has proved a thorn in the flesh of theoreticians ever since its 'periodic' variations were discovered. More recent observations show that it does not, as was first thought, vary regularly, like a pulsar, but in a much more erratic fashion. And the optical identification of the source confuses the issue still further, since in this case the mass of the X-ray star seems too big for it to be a white dwarf or a neutron star. We shall see the consequences of this in the next chapter.

But by and large X-ray astronomy is making some sense—even to the extent that Sco X-1 itself can be fitted in to the pattern. Oddly, one of the most remarkable developments in this respect was made by radio astronomers, who, early in September of 1972 took a quick look at Cyg X-3 while waiting for another more interesting radio source to rise above the horizon. To their astonishment they found that its radio strength had increased so dramatically that it had to be counted as a radio 'supernova'—but without any corresponding flare-up at X-ray or optical frequencies. The outburst was soon being monitored by observatories around the world, and the results of their observations were published in a special issue of the journal *Nature Physical Science*. It turned out that the best explanation for this astonishing outburst was very similar to the general explanation of how quasars 'work', but appropriately scaled down: the model involves a 'bubble' of plasma and magnetic field expanding at a sizeable fraction of the speed of light. One intriguing possibility is that, after a few hundred or thousand years, Cyg X-3 will end up with the double structure typical of radio galaxies and quasars—and, as we saw in the last chapter, of the radio emission of Sco X-1. Did the archetypal X-ray source Sco X-1 produce a radio supernova of its own sometime in the past? No one can be sure, but the possibility is attractive, and helps to fit Sco X-1 into the general pattern.

A great deal remains to be done before astronomers can claim to have a detailed understanding of X-ray sources. Graphic models have been presented, including one which describes the X-ray emission from such a source as coming from a mound of gas at the magnetic poles, about as high as the Post Office Tower and covering an area nearly as big as Hyde Park. But this, of course, is only an early step on the long road to discovering what makes these energetic sources tick.

The new wave of X-ray astronomy (post-Uhuru) has brought home to astronomers that they do not have to look to the far ends of the universe, to quasars and the like, in order to see extreme events—great explosions, radio supernovae like mini-quasars, and so on. Sudden and violent events are important within our own galaxy, as well as on a far grander scale. A better understanding of the violence which goes on in our galaxy will, it is hoped, help understanding of the violent events seen in extragalactic objects. Indeed, according to one interpretation of the radiation from Cygnus X-1, we have in our own celestial back yard at least one object as remarkable and extreme as anything anywhere in the universe—a black hole.

11 Black Holes & White Holes

A lot of rubbish has been written and broadcast about black holes over the past couple of years. It is hardly surprising that black holes should stir people's imagination; what is surprising is that the idea has taken so long to catch on. Black holes, as such, have been known about as an interesting theory for quite a long time—in fact, you don't even need the complexities of general relativity to find that black holes are required by theory. Newton's theory also requires that such oddities must exist. But come to that, there is no reason why a black hole should be all that odd—indeed, the universe in which we live might well be a sort of black hole in its own right. So why has there been all this fuss lately, and where has the fuss led to confusion?

First of all, the confusion. This is easily understood, because a lot of people (astronomers included) have lately been using the terms 'black hole' and 'singularity' as synonymous. They are not. A singularity is a peculiarity of Einsteinian (and other) theories, and singularities have no place in Newtonian physics. If a singularity, or something close to a singularity, existed in our own galaxy we might see some interesting results—and some people think that we can indeed see the evidence of such a local singularity in one of the recently discovered X-ray sources. Such a singularity is also a black hole. But a black hole need not be singular; in that case, what exactly is such an object?

Put at its simplest, a black hole is a region of space from which the escape velocity is greater than the speed of light. Most people today know about escape velocity from hearing and reading about space rockets. The strength of the Earth's gravity, for example, is such that any rocket, bullet or cricket ball, say, thrown upwards with a speed of less than 11 km per sec, must be pulled back to Earth. This can be expressed neatly in energy terms: if the kinetic energy that a particle has through its motion is less than the potential energy of the gravity field acting on the particle, then it cannot escape. But once this escape velocity is reached, any particle thrown outwards from the Earth can keep on going. Or at least, it can escape from the Earth. A rocket at the same distance from the Sun as the Earth would need a velocity of 41 km per sec to escape from the greater gravitational pull of the Sun and leave the Solar System. Because the Moon's gravity is weaker than that of the Earth, escape velocity is much less, and that is why Apollo lunar modules could manage with such small rockets.

Now, it's pretty obvious that the gravity of an object like the Sun is more than the gravity of the Earth, and so the escape velocity is higher. In fact, when you measure or calculate the escape velocity

Left: Stephan's Quintet of galaxies

needed at the surface of an astronomical body like the Earth or the Sun, both its mass and its density are important factors. The mass, obviously, because mass is what produces the gravitational field our rocket has to escape from; and the density because the more dense an object is then the closer its surface is to its centre. Since the strength of gravity depends inversely on the square of the distance over which it is operating, it is obviously going to be harder for our rocket to escape from the surface of the Sun as it is today, than it will be in the distant future, when the Sun becomes a giant and expands out beyond the orbit of the Earth.

So how do we get a black hole? First, we can ignore density and just keep on increasing the mass of the body our rocket has to escape from. As the mass goes up, so does the escape velocity, and in a surprisingly short time we find that the speed of light is needed to escape from the gravitational field we have produced. This is true whatever theory of physics or mathematics we use. But Einstein's theory tells us, and experiments confirm, that *nothing* (except tachyons!) can travel faster than light. So if we have an object so massive that light cannot escape from it, then nothing can escape, and the inside of the black hole is almost a self-contained universe. Almost, but not quite, since, of course, things can still get in from outside, even if nothing inside can ever get out.

It is quite reasonable, incidentally, that light should be affected by the escape velocity restriction, even though the velocity of light is always constant, because photons—particles of light—have only a certain amount of kinetic energy, just like other particles. There is one difference in behaviour between photons and other particles, when climbing out of a gravity field (or gravitational 'potential well', as it is graphically called): whereas an ordinary object like our rocket loses speed as it escapes (trading kinetic energy for potential energy), a light particle, or beam, becomes redshifted. This again corresponds to a loss of energy, and in the case of a black hole the redshift becomes infinite, so that the light cannot escape. This gravitational redshift has been measured, and is another confirmation of the accuracy of Einstein's general theory of relativity. At one time, indeed, there was a suspicion among astronomers that the large redshifts seen in quasar spectra might be gravitational—but that explanation no longer really fits all the available observations.

But there is another way to approach the creation of a black hole, apart from increasing the mass of an object enormously. If the density is so important, can we get a black hole by squeezing some ordinary object down small enough? The answer is that in principle we can— but it might prove rather tricky in practice.

In terms of relativity theory, a black hole corresponds to a region of space that is curved so much that it closes in on itself under the force of gravity. It turns out that within the framework of relativity theory there should indeed be, for any mass, some critical density which will produce a black hole. In other words, if the mass is squeezed within a critical radius—the Schwarzschild radius, so named after one of the

Overleaf: Left—Copernicus being prepared for launch. Right—The launch of the Copernicus satellite

great mathematicians—it will become a black hole. This radius is easily calculated—it is just twice the product of the gravitational constant and the mass, divided by the square of the speed of light. (As it happens, this is exactly the same as the equivalent black hole radius in Newtonian physics.)

As you can guess, for any object of reasonable mass the Schwarzschild radius is ludicrously small, in everyday terms, and the density of the resulting black hole is ludicrously big. Even the Sun would have to be squeezed into a ball only 6 km across before it became a black hole, and its density would then be 2×10^{16} gm/cm^3 (that is, 20,000,000,000,000,000 grams per cubic centimetre)! But if we imagine bigger and bigger masses turned into black holes, the figures become more reasonable, although quite unwieldy in human terms. A quasar, for instance, with a mass of about three million million Suns could be made into a black hole one light year (about nine and a half million million kilometres) in radius, and its density would be only two thousand-millionths of a gram per cubic centimetre (2×10^{-9} gm/cm^3).

That is quite a reasonable density, in astronomical terms, and it begins to look as if black holes might exist, even though they cannot be made by man. What if we go to the ultimate limit, and consider our whole universe? The best thing to do here is switch the argument around slightly, and ask: if the universe has a radius of about ten thousand million light years, how dense would it have to be in order to be a black hole? The answer is surprising. A density of only 10^{-29} gm/cm^3 will do the trick. Now that is pretty small; so just what is the density of the universe? The answer is we don't know for sure; but by averaging out all the visible bright galaxies astronomers have reckoned that the average density must be about 10^{-30} gm/cm^3, which is just one tenth of the amount needed to make the universe a black hole. Of course, it's not really correct to talk of the universe as a black hole, since where could light 'leaving the universe' be going anyway? In some sense, any 'universe' must by definition be a sort of black hole. Cosmologists have, however, a great interest in a problem which is superficially similar. They ask the question: 'Is the universe open or closed?' We shall see what they mean by this, and what answers are possible, in Chapter 12. In the meantime, it's worth remembering that there could well be enough dark matter in the universe, not visible as bright stars in galaxies, for the critical density of 10^{-29} gm/cm^3 to be reached.

So what have these simple examples told us about black holes? Merely that if you are going to make one with a mass much less than that of a galaxy or quasar, something very odd must happen to produce the enormous density required. And that explains the recent interest and excitement caused by black holes; some people believe that 'something very odd' has indeed happened to the X-ray source Cyg X-1, and that it contains a black hole with a mass only a few times that of the Sun.

Before we look at the evidence for and against this belief, we should

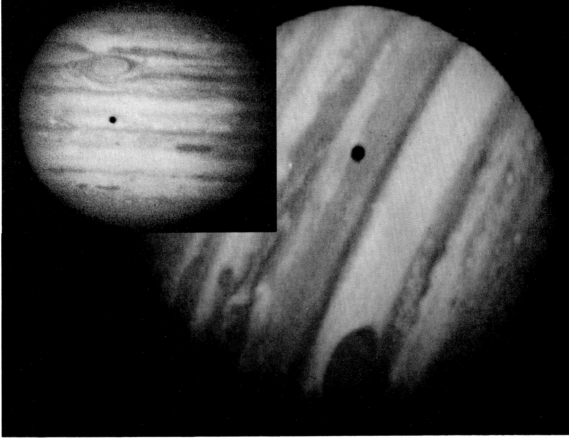

Left: A rich cluster of galaxies in the constellation Hydra, photographed by the Schmidt telescope shown on page 100

Overleaf: Lightning over Kitt Peak Observatory, Arizona

see what kind of 'very odd' effect could, in theory, produce such a phenomenon. We saw in Chapter 4 how astronomers construct 'models' of stars by considering the equations of physics which govern the behaviour of all matter. An ordinary star is held up against the inward pull of gravity by the thermal energy of its interior—the hot plasma of atomic nuclei and electrons pushing outwards. Once a star becomes old and uses up all the nuclear fuel in its interior, it shrinks down because gravity is no longer balanced by the decreasing thermal pressure. Eventually, a stable situation can be reached when the star becomes solid; all the atomic nuclei are packed together in a crystalline array. The star has become a white dwarf, and can be likened to a single giant, cold crystal. This is the path our own Sun will eventually follow. Such a white dwarf is stable because now the inward pull of gravity is balanced not by thermal pressure, but by the electric forces repelling all the positively charged nuclei in the crystal lattice one from another.

But even that need not be the end of the story. If more and more matter is added to a star, gravity can be increased indefinitely. But the forces of electrical repulsion holding a white dwarf up are only limited. If a star is massive enough, these forces can in turn be overcome.

In fact, a star only a little more massive than our Sun will not form a stable white dwarf, but will be squeezed by its own gravity into an even more compact form—a neutron star. What happens, in effect, is that all the electrons and protons of the original plasma are squeezed together into neutrons, and the neutrons are squeezed together cheek by jowl to form what amounts to one giant 'atomic nucleus', as massive as a star but only a few kilometres across.

This was predicted by general relativity some time ago; the discovery of pulsars, and the evidence which shows that pulsars are very probably neutron stars, has been taken as a striking proof that general relativity does indeed 'work' as a description of the physics of the universe under such extreme conditions. And there is still another stage to this prediction; given a little more matter still, even a neutron star cannot hold itself up against the crushing force of gravity. There is no room left for neutrons to move closer together, but according to general relativity even these 'fundamental' particles cannot withstand the all-powerful force of gravity.

If a cold star (or anything else) has enough mass, the neutrons will be destroyed and all the mass crushed down into a mathematical point—a singularity. This is a most disturbing concept, and mathematicians are still arguing about whether or not this can really happen, or whether the prediction shows that even general relativity theory is not any good at describing the most extreme conditions of all. Perhaps there is some sort of repulsion force which switches in to stop the collapse to a singularity. That is a matter for the relativists to argue about. What concerns us here is that if this gravitational collapse really does occur, then obviously the collapsing star must pass through its Schwarzschild radius—however small that may be—

on the way to becoming a singularity. In other words, if general relativity is correct, any star with a mass slightly greater than the Sun's must eventually become a super-dense black hole. And that is the kind of black hole which has caused so much excitement recently.

The argument which led some astronomers to believe that Cyg-X-1 might be a black hole of this kind—a singular black hole—is pretty easy to follow. Because Cyg X-1 is identified with an optically visible star which can be studied with ordinary telescopes, it is possible to work out its mass in a slightly roundabout way.

The visible star cannot itself be the X-ray source because, as we saw, the X-rays must come from the gravitational energy of matter falling on to a small, dense object—a white dwarf, a neutron star, or, it seems, a black hole. What we can see with telescopes is the companion star, from which all the matter is being removed by tidal action. And that star is a member of a family well known to astronomers; it is a B-type super-giant, with a mass equal to twenty or more Suns.

This is where the fun begins. Knowing roughly the mass of the giant, and knowing the period of the binary pair from the periodic variations of Cyg X-1, it is not too difficult to work out the mass of the companion, the X-ray source itself. The answer is, about eleven times the mass of the Sun—and that is far and away too much for a white dwarf or neutron star. So some astronomers got rather excited during 1973, and believed that a black hole had at last been identified.

Alas, it's not quite as simple as that. Although the periodic variations of Cyg X-1 tell us that there must be more than one star in the system, *there is nothing which indicates that there can be only two stars there.* Optical astronomers know that about a third of all known 'binary' systems actually contain at least three—and sometimes more —stars. What the periodic variations tell us is that Cyg X-1 is a *multiple* system; they do not say that the appropriate multiple is two rather than three.

This possibility seems to have occurred to astronomers in the USA and Britain at about the same time, and in the spring of 1974 two groups independently suggested how Cyg X-1 could be explained as a triple system. In the simplest case, we can say that there is indeed a main (primary) star of about twenty solar masses—we have to say that, because we can see that star! But it's quite permissible to say that the secondary, adding up to eleven Suns in all, is made up of a hot star of ten solar masses, which is orbited in turn by a neutron star of about the same mass as our Sun. Now we can explain the periodic variations, and the production of X-rays, without invoking black holes at all.

So the excitement was a bit premature. Singular black holes might exist (we shall see another way they might form in Chapter 13), and indeed there might be one in Cyg X-1. But a good rule of thumb in astronomy (and in everyday life) is that if two explanations can both

Overleaf: Left—The central region of M8 showing the 'Hourglass' Nebula. Right— Two pictures of the Space Laboratory which will be carried into orbit by the Space Shuttle in the 1980s

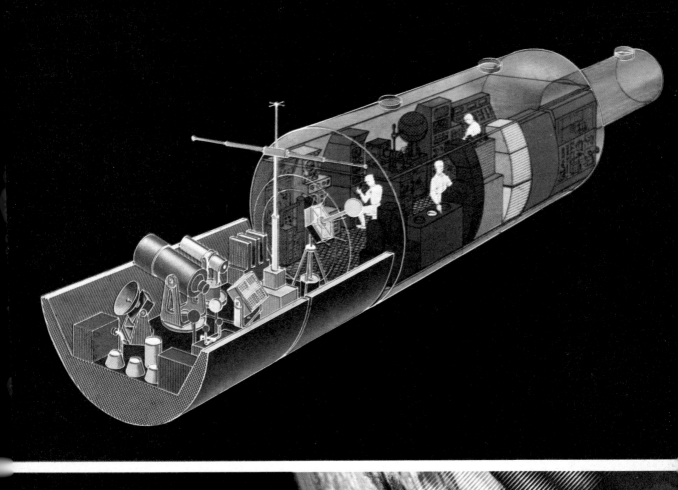

account for the same effect, then the simplest explanation is most probably correct. This rule, by the way, is known as 'Occam's razor': in the case of Cyg X-1 it is simpler to explain the observations in terms of three kinds of star which are all known to exist than to insist on dragging in a new kind of astrophysical phenomenon. As yet, it seems, we cannot say whether or not general relativity really does hold good right down to the bitter end of gravitational collapse.

But there is a consolation to keep the more speculative astronomers amused while they wait for firm evidence one way or the other. A curious thing about the laws of physics is that they are reversible. On the grand scale, there is nothing to say that they can't run 'backwards'. This has interesting philosophical and cosmological consequences, of which more later. But in the case of gravitational collapse it means that by turning the equations around we can describe, in mathematical terms, a new kind of phenomenon. Instead of collapse, we have a sudden explosion of material outwards from a point. Logically enough, the relativists have dubbed this hypothetical phenomenon, the opposite of a black hole, a 'white hole'.

But is it so hypothetical? Is there anywhere in the universe that we see great quantities of energy being liberated from very energetic sources which seem to be exploding outwards into space? Yes indeed —quasars and the active centres of some galaxies behave in just this sort of fashion (Chapter 3). Now, I am not prepared to stick my neck out and say that quasars *are* white holes. But I have got just as much faith in that theory as in the theory that Cyg X-1 harbours a black hole. And the beauty of it is that it is going to be very difficult for anyone to prove that I am wrong.

Left: Centaurus A, a violently active radio galaxy

12 Gravitational Radiation

So the 'singular' kinds of black holes (but not the other variety) are still a prediction of general relativity which has yet to be confirmed. Decades ago, relativists realised that general relativity also predicted another phenomenon which could provide an 'ultimate' test of the theory's accuracy. This is the phenomenon of gravitational radiation.

According to general relativity, certain kinds of movement of gravitating bodies through space should produce gravitational radiation, in much the same way that movement of electrically charged particles causes electromagnetic radiation, under the right conditions. But this radiation should be very weak indeed if relativity theory is correct, and for a long time gravitational radiation was just another hypothetical toy of the mathematicians. Then, through the intensive work of one experimenter, it looked briefly as if gravitational radiation had been detected. As it turned out, those particular experiments were probably wrong after all. But the apparent discovery of the radiation aroused so much interest that many experimenters are now tackling the problem of building gravity-wave detectors, and it seems that the real discovery of gravitational radiation may now be just around the corner.

When Joseph Weber first claimed to have detected gravitational pulses coming from the region of the centre of our Galaxy, the relativists were quite disconcerted. Initially, they were pleased to have confirmation that general relativity's prediction had come true—but then they realised that one could have too much of a good thing! The 'Weber pulses' were so strong and so frequent that, if they were gravitational radiation, general relativity must be seriously wrong in predicting that only weak pulses would exist in our galaxy! But just what were the experiments which caused this disconcerting situation?

Although Professor Joseph Weber had been working towards the detection of gravitational radiation for many years, it was only as recently as 1969 that he produced claims of success which were, at first, widely accepted by astronomers. This was the trigger for the recent wave of excitement about gravitational radiation. The experiment which forms the basis of Weber's lifework is designed around two massive aluminium cylinders—each about five feet long, two feet in diameter, and weighing about one and a half tons—which are placed 1000 km apart, one at the Argonne National Laboratory and the other at the University of Maryland. The theory is that pulses of gravitational radiation should gently shake them; or more exactly, that gravitational radiation literally stretches and squeezes the cylinders as a wave passes, in a way vaguely similar to a sound wave causing the air to expand and contract as it passes.

With modern electronic techniques, it is not too difficult to measure very small compressions and stretches of such long cylinders; Weber has used piezoelectric (electricity produced by pressure) techniques, in which certain kinds of crystal attached to the test bar are squeezed and as a result produce a small electric current. With this system, a change in length of one hundred-million-millionth of a centimetre (10^{-14} cm) can be detected.

There has been a lot of argument among relativists lately about whether or not that is good enough to detect the kind of very weak gravitational pulses which, according to general relativity, might be expected to occur in our galaxy. But there is also another problem.

Such very sensitive detectors will clearly also detect lots of other things, unless elaborate precautions are taken. The bars must be screened, in case disturbances of the Earth's magnetic field affect them, and they must be supported, as far as possible, in such a way that earth tremors (or even heavy lorries driving past!) cannot shake the equipment.

Of course, that is impossible. So instead two detectors are used, as in Weber's experiments. A lorry driving by in Maryland won't affect the Argonne detector, so Weber's team look for coincidences between the two sets of data, to find gravity pulses. Even then life is not all plain sailing, and there are certain mathematical and statistical ways of choosing coincidences which are most likely to be genuine. But by and large it is pretty clear that Weber's detectors were (and are) detecting something. Clear enough, at any rate, to spur on work in many laboratories around the world, some trying to duplicate Weber's experiments, as a check, and others using rather different equipment and hoping to find out different things about the supposed gravitational radiation. Even radio astronomers got in on the act, looking for radio pulses which might accompany the gravity pulses. But no one has had any luck.

By 1972, few of these new experiments were complete. But the theoreticians were already rather concerned that in the short period since 1969 Weber had detected too much gravitational radiation. The implications of Weber's claims were discussed at a meeting of the Royal Astronomical Society in January, 1972; and there it was pointed out just what kind of astronomical process would be needed to produce gravity pulses that would fit Weber's observations.

Basically, something pretty violent and energetic is needed to produce pulses of gravity waves that can travel across the galaxy and be recorded on Earth. A supernova explosion, perhaps, or the collapse of an object into a black hole, or even a collision between black holes. Now, these just are not everyday events. Yet Weber has now for several years been detecting one or two pulses a day, at least.

Put another way, the problem can be seen in terms of mass and energy. The two are equivalent, of course, and related by the well known equation $E = mc^2$. So it is fairly simple to work out how much mass is being turned into energy each day to produce Weber's alleged

Overleaf: Gravitational radiation detection equipment at Glasgow University

115

gravity pulses. The answer implies that our galaxy is 'losing' the equivalent of twenty thousand Suns each year—and that is such a high rate that it ought to produce clearly visible effects in many other ways.

So there seemed to be two possibilities: either Weber is not detecting gravitational radiation, or the radiation is produced in some less extreme manner, which would mean that general relativity is incorrect. By the middle of 1973 a clear consensus had emerged from the deliberations of the relativists and the tests of the experimenters. Weber's apparatus is detecting *something*—but not pulses of gravitational radiation.

Probably the first occasion on which this consensus became clearly apparent was at a joint Oxford-Cambridge-London meeting to discuss the problem, held at All Souls College, Oxford, in April 1973. Even then—and, indeed, even today—some people could not bring themselves to dismiss completely the idea that Weber had detected gravity waves. But the latest evidence has convinced the majority.

Several teams of experimenters have now failed to duplicate Weber's observations. These include a group in the Soviet Union and a joint German/Italian experiment with detectors in Munich and Frascati; the latter is particularly interesting because it is designed to be as nearly as possible a duplicate of Weber's Argonne/Maryland experiment. In Britain, a more sensitive detector is being built in Bristol, and a detector rather smaller than Weber's, but with just about the same sensitivity, is being run at the University of Glasgow with consistently negative results. And the story is much the same when other experimenters in the USA try to detect gravity waves. Whether the apparatus is large or small, in Europe or in America, with sophisticated 'signal processing' statistics or without, no one can detect anything which they are prepared to offer as one genuine example of a gravity pulse. And still Weber is recording one or more pulses a day—so whatever he has been detecting is still there, and has not 'switched off' since the new experiments were built.

But the theoreticians at the Oxford meeting of April 1973 were far from displeased by these negative results. According to their calculations, even a 'sensible' event like a supernova explosion could not be detected by equipment with the sensitivity of Weber's apparatus. To detect supernovae, in any case, there is no point in just 'looking' at our own galaxy, because we only have one or two supernovae each century. What is needed is an experiment sensitive enough to record gravity pulses from supernova explosions in other galaxies, like those of the Virgo cluster. And that would need equipment a hundred million times more sensitive than Weber's.

Looking at the question another way, the theoreticians at the Oxford meeting agreed that they would only begin to worry about relativity theory being wrong if no gravity pulses could be detected with equipment even a million million (10^{12}) times more sensitive than that being built today! That might seem rather to have killed off the young science of gravitational wave experimentation! But Dr R. W. P. Drever of the University of Glasgow thinks that the first improvement—by a factor of a hundred million in sensitivity—is possible within a few years.

This will involve even more exotic experiments. The ultimate refinement of Weber's approach would be a five or six ton cylinder levitated by superconducting magnetic coils and cooled to nearly zero degrees absolute to reduce the thermal motions of its constituent atoms. Alternatively, space probes could be used to establish a very long 'baseline', with distances, and hence oscillations, monitored by laser beams.

The fascinating thing about these science-fiction-like ideas is that if theoreticians had been left to their own devices no one would have dreamed of such exotic experiments becoming a reality for decades. Because Weber pressed on when theory said he must be unsuccessful, and found something which frightened other people into frenzied activity, great advances in technique have already been made. Now, the road is clearly enough laid out for detection of real gravity pulses to be very likely within a matter of years. And this would not have been possible without Weber's work.

But just what is Weber detecting? Several suggestions have been made, but the one which I favour (for no particular scientific reason, I just like the idea) is that it is some sort of magnetic effect. Weber's equipment does certainly contain magnetic components, and it is very difficult to screen out all outside magnetic fields. One analysis, carried out at the Bell Laboratories, has found a tie-up between the times at which Weber's detectors actually detect something, the times of certain changes in terrestrial magnetic activity in the ionosphere, and the Sun's activity. The statistical significance of the correlations is not very great, as such statistics are measured, but it provides food for thought. During the same analysis, the Bell Laboratories group also found a hint of a correlation between sunspots and earthquake activity. That is interesting on two counts: first, perhaps the solar activity affects the earthquakes which in turn might affect both Weber's detectors even though they are 1000 km apart; secondly, the idea that sunspots might influence earthquake activity is something that I have been studying myself for some time, quite independently of any connection with either Weber's work or the Bell Laboratories study. But that is another story (see *The Jupiter Effect*, by John Gribbin and Stephen Plagemann, Macmillan, 1974).

As things stand today, then, it still looks as if Einstein's theory of relativity provides the best description we have of our universe. This, of course, means much more than whether or not black holes and gravity waves exist—and this seems the right place to look at the universe as a whole, and our understanding of it. In other words, the science of cosmology.

13 The Cosmic Background Radiation

No discussion of cosmology would be complete without mention of the cosmic background radiation. The discovery and identification of this faint background noise in the electromagnetic spectrum ranks as one of the most significant developments in the history of science. Curiously, however, whereas pulsars, quasars and X-ray sources seem to grip the popular imagination, the background radiation has largely gone unsung. It may seem relatively unexciting in its present form – a weak hiss of radio noise detectable at frequencies between 1 MHz and 500 MHz, but this is the very echo of the big bang in which the universe as we know it began—the cosmic fireball.

This background radiation was first discovered in 1965, and given its name because it seems to be coming from all directions of space. Wherever radio astronomers point their antennae, they get the same underlying hiss of background radiation. The other name for this radiation, the '3 K blackbody radiation', describes the nature of the noise. An ideal radiator of energy is called a 'black body', and well known laws relate the amount of energy emitted at any one frequency to the temperature of the black body. First measurements of the cosmic background radiation showed that it is equivalent to radiation from a black body with a temperature close to 3 K, that is −270 °C. Later estimates refined this measurement, and now we can give the black body temperature equivalent of the background radiation more precisely, as about 2·75 K. That sounds pretty feeble, and it is hardly surprising that such a discovery has less immediate impact for the non-astronomer than the discovery of quasars, pumping out their vast quantities of energy into space. But, unlike quasars, the black body background seems to be *everywhere* in the universe, and even for a radiation temperature of 2·75 K that adds up to a lot of radiation. Einstein's famous equation relates mass and energy and, by adding up all the energy in the black body background, it has been calculated that the mass equivalent of all this radiation is roughly the same as the mass of all the known galaxies and quasars added together. In universal terms, this radiation is as important as all the matter visible in brightly radiating objects.

So where did all this energy come from? Again, we can compare the situation with that of the visible matter in the universe. The matter, of course, is concentrated in lumps—stars and galaxies. But if it were all spread out to fill space uniformly, as the background radiation does, the density would be only 10^{-30} gm/cm^3, give or take a factor of 10. In other words, the matter would then look as unimpressive as the background radiation. But the universe is expanding; clusters of galaxies are moving further apart from one another, and the density of matter is falling. The equivalent effect on the background radiation is to reduce its black body temperature, which is equivalent to reducing the radiation density. What then were conditions like in the distant past? We can imagine one possible answer to this question by considering what would happen to the universe we see about us if, instead of expanding, it contracted by a very large amount.

When the universe was smaller, both matter density and radiation density must have been greater. If, in our imaginations, we run the 'film' of the evolution of the universe backwards, we first see clusters of galaxies squeezing together, then the very structure of stars and galaxies being destroyed, as all matter is squeezed into one single lump of primeval matter. At the same time, the black body temperature of the background radiation rises; soon the energy density of the radiation is so great that it plays an important part in interactions with the elementary particles which are now all that remains of the once complex structure of matter.

By making these kinds of extrapolation backwards in time, cosmologists have come up with a very plausible picture of how the universe may have evolved from such a hot primeval fireball into its present state. (How it ever got to be a primeval fireball, however, is another story, as we shall see in the next chapter.) This evolutionary development is described something along these lines: at the beginning, when the temperature of the fireball universe was around 10^{12} K, energetic particles, antiparticles and radiation were interacting continuously in a violent maelstrom of reactions. Particles would frequently be annihilated and turned into radiation, while photons of the radiation would often be converted into pairs or showers of particles. As this maelstrom expanded, however, the temperature would have dropped and the situation become more orderly. At 10^{11} K, the more exotic elementary particles were no longer being created out of the radiation; they would have all been eliminated in annihilation processes, leaving a universe full of protons, neutrons, electrons and positrons (the latter still involved with annihilation and production reactions involving the radiation). Neutrinos would have been produced very early in the fireball, and are presumably still around somewhere; but in view of the problems encountered in searching simply for solar neutrinos, it looks as if it will be a long while yet before anyone can have a hope of detecting these primeval 'background' neutrinos.

At 10^9 K, even production of further electron-positron pairs required too much energy for the weakening background radiation, and most of the pairs would have annihilated themselves. For some reason, however, there seem to have been a few extra electrons, over and above the number of positrons in the universe. These 'spare' electrons were of critical importance to the development of the universe we know—indeed, all our surroundings (and ourselves) are

Bell Laboratory's horn antenna at Holmdell, New Jersey, which was used by Penzias (inset right) and Wilson (inset left) in early studies of the cosmic background radiation

in a very real sense the product of this minor asymmetry back in the cosmic fireball. Also at around 10^9 K, many protons and neutrons were 'cooked' into helium (in a similar way to the nucleosynthesis that now goes on in the Sun, described in Chapter 8); but not until the temperature dropped to 5000 K would the positively charged protons and helium nuclei combine with electrons to form neutral atoms, of hydrogen and helium. Once the electrons and the atomic nuclei were bound in electrically *neutral* atoms, radiation could no longer interact with the matter to any significant degree, because such interactions involve the acceleration of *charged* particles (or, at much higher energies, the creation and annihilation of pairs of such particles). At this stage, the photons lost all real contact with the atoms, radiation was decoupled from the matter, and we had the beginnings of the conditions which lead to the production of stars and galaxies. Meanwhile the background radiation, left to its own devices, got weaker and weaker, until now we see it as a mere 2·75 K whimper of its former self.

It's worth noting, perhaps, that the surface of the Sun today is at about 5000 K, and that conditions there are very similar to the state of the whole universe at the time that matter and radiation decoupled.

So the presence of the faint echo of the cosmic fireball which astronomers can detect today, coming uniformly from all directions in space, gives them very precise information about how the universe got into its present state. On the other hand, there is still plenty of scope for cosmologists to speculate about how the cosmic fireball itself came about, and to ponder on where the universe is 'going'.

14 Cosmology

Cosmology is a curious science which inhabits the borderlands between mathematics, philosophy, and even religion. In terms of observations, cosmology has only been an 'experimental' science in a big way for the past few decades. But even the ancient Chinese 3000 years ago had thoughts about the nature of the universe and man's place in it, and much of the work of Newton and his contemporaries is also recognisably within the meaning of the term cosmology. Put at its simplest, cosmology is just the study of the universe as a whole, where it 'came from' and where it might be 'going'. So if the best theory of the universe you have is that the Earth is a shallow saucer riding on the back of a great turtle, then theories about the shape and size of the saucer, the means by which the Sun moves overhead, and even the name of the turtle would certainly count as cosmology.

Here and now I am supposed to be restricting myself to describing the new astronomy, and there have indeed been remarkable recent developments in cosmology which are well worth describing. But since most of modern cosmology is built on Einstein's general theory of relativity, and since that kind of modern cosmology began only some fifty to sixty years ago, it is perhaps worth sketching in the outline of how the science got from that point to where it is today, before we look at some of the more exotic modern ideas.

The most important piece of observational evidence in cosmology is the redshift of light from distant galaxies. Because the universe as a whole is so big, each galaxy in it corresponds to one 'particle' as far as cosmologists are concerned. Our own galaxy (the Milky Way) is made up of something like a hundred thousand million (10^{11}) stars, and it is by no means the largest. Even so, when we are thinking in cosmological terms, such congregations of stars count simply as one particle; just as we treat the planets as points, or single particles when we are calculating their orbits around the Sun, even though we know that they are made up of many millions of atoms and molecules.

It was not until the 1920s, however, that most astronomers came to accept that there are many galaxies besides our own Milky Way, and that our galaxy is no more significant on a universal scale than the Sun is on a galactic scale. Edwin Hubble, an American observational astronomer who was one of the great pioneers of the twentieth century, settled the issue in the mid-1920s when he was able, using the 100-inch telescope at Mount Wilson, to detect certain types of star, known as 'Cepheid variables', in some of the spiral nebulae.

The importance of this work lay in the fact that the rate of visible variation (in brightness) of Cepheids is dependent on the 'absolute' brightness of the star. So Hubble could, by measuring the period of these variations, work out how bright the stars must be, and therefore how far away they must be in order to appear as they do to an observer on Earth. Since Hubble's pioneering work, the distance estimates have been greatly improved. But the essential point remains —the distances calculated in this way are far greater than the diameter of our galaxy. In other words, many of the so-called nebulae are collections of stars outside our galaxy—which form whole galaxies in their own right.

The snag about measuring distances from Cepheid variables is that these are not the brightest of stars, and so the method only works for nearby galaxies. Hubble extended his measurements of the distances of galaxies by assuming that the very brightest super-giant stars in all galaxies have the same 'absolute' luminosity. This could be checked for the nearby galaxies—the local group—from the Cepheid method, and it worked fairly well. This pushed out the measured distances of galaxies to ten million light years—but even that is too small a distance for cosmological work. What about the galaxies which are so far away that they cannot be resolved into individual stars?

Hubble did press further by assuming that all galaxies are equally bright, so that the dimmest galaxies seen from Earth are simply the furthest away. That gave him rough measurements out to 500 million

Below: The Small Magellanic Cloud

light years; but, as more recent studies have shown, galaxies do vary considerably in brightness, and Hubble's estimates are not accurate at the larger distances. Without some reliable distance indicator, cosmology could not have developed as an observational science, but would have remained the province of philosophers and theoretical mathematicians alone. That is where the redshift comes in.

As we saw in Chapter 3 (which was rather putting the quasar cart before the galaxy horse!), the redshift of features in the spectra of galaxies indicates that they are receding with velocities proportional to their distances from us, on average. This discovery depended, of course, on measuring the distances to the nearest galaxies, so that the velocity/distance relation (Hubble's law) could be formulated. But once the law was known, it could be used to calculate the distance of any visible galaxy, by measuring its redshift on the spectrum. With large redshifts (of greater than 2), the velocity of the receding light source is approaching that of light, and relativity effects must be taken into account: the simple Hubble's law no longer works. But the principle is the same: greater redshift implies greater velocity and greater distance. Of course, there are random movements of galaxies which affect the Doppler redshift. But for distant galaxies the cosmological effect is of overwhelming importance.

With Hubble's law, and the advent of the 200-inch telescope some thirty years ago, cosmologists were at last able to obtain experimental —that is, observational—evidence accurate enough, and covering a large enough fraction of the universe, for different cosmological theories and models to be tested—to see how well they explained the nature of the universe in which we live.

One curious historical development came about, in fact, partly because the expansion of the universe was not known during the first two decades of this century. Einstein found that although his relativity equations could be used to describe the behaviour of model universes of various kinds, it was very difficult to produce a 'static' model, in which the basic particles stayed in more-or-less the same situation. The relativity models always seemed to want to expand or contract, and a stable situation could only be created by bringing an extra term into the equations—a 'cosmological constant'.

At the time, that seemed to be a fundamental flaw in relativity. But when the expansion of the universe was discovered, and Hubble formulated his 'law' of the recession of galaxies, astronomers quickly realised that in a sense this had been predicted by relativity theory. An expanding universe can be neatly explained without any need for the cosmological constant.

That has not, however, stopped various people from invoking different cosmological constants to produce different kinds of theoretical model universe. Indeed, some cosmologists have chosen to use theories other than Einstein's to calculate their models. But by and large the more straightforward applications of relativity theory are quite good enough to explain the observed properties of our universe (indeed, strange though it may seem, even Newtonian

Left: The open spiral galaxy
M101

mechanics can do a good job of describing the universe). Perhaps future observations will show the need for other theories to describe our universe. But for the present, even sticking within the bounds of relativity theory we can find plenty of food for thought.

Basically, an expanding universe must be some phase in the development of one of three kinds of universe. Such a universe might continue expanding forever, or it might expand to some definite limit and stop, or it might eventually 'turn around' and begin to contract. As with a black hole, the factor which decides what happens is the gravitational attraction of everything in the universe. If the universe is dense enough, then it is the equivalent, if you like, of a black hole from which nothing can escape. In that case, the recession of galaxies will slow down and eventually be reversed, with galaxies rushing towards each other and eventually producing a singularity.

That idea is very attractive, because the expansion observed today seems to indicate that everything must have started out from just such a singularity. Perhaps, say some cosmologists, there is a cycle of expansion and contraction, with collapse being reversed again at the singularity and turned into another wave of expansion.

But if the density of the universe today is rather less, we could have a critical situation in which the expansion was just cancelled out, stopped, without being tipped over into contraction. That is the least satisfying theory of all, because why should there be only enough mass for that to happen, no more and no less?

So the only basic alternative to the cyclic model that has any appeal is the model of a permanently expanding universe, which contains too little mass for its gravity ever to halt the expansion. This means that the universe is always changing, or evolving. Some cosmologists do not like that idea, which implies a definite beginning, if not a definite end; so they suggested the 'steady state' theory, in which new matter is always being created to fill the gaps as the galaxies move apart. The steady state idea cannot be disproved; basically, it is a philosophical question whether you prefer 'once and for all' creation, in what is known as the 'big bang', or continuous creation. But to my mind the first alternative is preferable. Although I do not agree with those astronomers who argue that the steady state theory is dead, I shall not describe it in detail simply because there seems no need for it. It seems to me that, like so many other cosmologies, continuous creation is an entertaining mathematical exercise which probably has nothing much to do with our universe (unless, perhaps, quasars really are 'white holes' where matter is created!).

There are far too many interesting cosmologies to describe in detail here. So I shall choose just two ideas to represent modern theories about our changing universe, one a straightforward cyclical model and one a novel explanation of the 'big bang' singularity.

The cyclic, or closed, models neatly sidestep two awkward questions: 'Where did the universe come from?' and 'Where is it going?' All you have to do is say that the collapse phase somehow (essentially by magic!) produces a 'bounce' at the singularity and, *voilà*!, we have, in each cycle, an expanding universe just like the one we see around us. By averaging out the two halves of the cycle we can even get a kind of 'steady state' situation. The main snag with the cyclic idea is that we cannot see enough matter around us in the universe to fit the theory. To make our universe a 'closed' one, what is needed—according to the best modern estimates—is between four and ten times more than the 10^{-30} gm/cm^3 or so which is found in visible bright galaxies. But it is quite plausible to argue that black holes—whole dead galaxies of black holes, if you like—could supply the 'missing mass'. Most cyclic models have little appeal beyond this. But one idea put forward in the early 1970s by Dr Paul Davies neatly explains a whole host of physical puzzles in such a model.

First, there is the problem of the 'arrow of time'. The universe is expanding, stars as well as people grow old and die, and things are just generally wearing out. This thermodynamic, or statistical, running down of everything on a large scale conflicts with the fact that most physical laws are reversible—they have no inbuilt direction to say which way is 'the future' and they would work equally well if everything were reversed. Then, stars would grow younger, matter would emerge from white holes instead of disappearing into black holes, and so on. In effect, Davies says 'why not reverse the arrow of time in next-door segments of the universe's cyclical life?'. In that way, everything would be recharged between cycles, and the philosophical problem of

Below: The Lick Observatory in California, viewed from the east

Above: Spiral galaxy NGC253,
seen almost edge on

the arrow of time would be resolved.

There is also the question of antimatter. There is no physical reason why our universe should be composed just of matter as we know it, and indeed there may be whole galaxies of mirror image antimatter. Matter and antimatter ought to have an equal chance of being made in a big bang. But Davies has an alternative suggestion here too—in the next door cycles of the universe, matter can be replaced by antimatter, and so on.

Finally, in Davies's model radiation in our cycle of the universe is converted to matter in the next cycle, and vice versa.

When all these bits are assembled mathematically, the model is very neat. At the end of a cycle, matter and radiation are scrambled up in a cosmic fireball, which becomes the big bang for the next cycle. In that cycle, radiation and antimatter follow a pattern of expansion and contraction with a reversed arrow of time. The next cycle would then be the same as the first—indeed, they would be mathematically indistinguishable, and Davies suggests that instead of an infinite pattern of cycles there are just two. The end of Cycle 1 is the beginning of Cycle 2 and the end of Cycle 2 is the beginning of Cycle 1.

This rather mind-boggling 'cosmic merry-go-round' idea does not seem to be taken too seriously by cosmologists. It is regarded as just a pretty way of putting the equations together. But it does work, it accounts for all the observed properties of the universe, and it explains the most baffling problems facing cosmologists.

With other theories, the question of the origin of the universe becomes paramount. There are many satisfactory enough ways of explaining what happens *after* the 'big bang'. They do not always agree with each other, and there is plenty to be worked out about how galaxies form, what quasars might be and so on. But the real question that falls within the province of cosmology rather than any other science is the question of the origin—the big bang itself.

The discovery of the microwave background radiation which fills the universe has provided almost incontrovertible proof that there really was a big bang in which the universe formed (or, at least, a singularity which might have been the switch-over point in a cyclic universe). And lately the cosmologists have waxed even more philosophical than usual in their attempts to account for the origin of the universe—or, turning the problem on its head, to account for our presence in such a universe as the one in which we live.

Some of the most fascinating work on this problem is by various cosmologists who attempt to bring quantum mechanical ideas into their theories—or philosophies. Quantum theory, for example, cannot predict exactly what will be the outcome of any experiment. What the theory does give is the probability of each possible outcome of an experiment. The conventional interpretation of this in philosophical terms is that the universe 'chooses' at random one of the many possible outcomes, and this random effect, operating on the scale of atoms and smaller particles, averages out to produce the unique universe we live in. But a science-fiction-like alternative has now been developed seriously. In this theory, all possible quantum alternatives are always produced when the universe is faced with such a choice. The entire universe is then continually splitting into myriads of 'parallel universes' in which all possible alternatives are realised. Since only a few such universes would be suitable for life, we can only exist to think about such problems in a universe very much like the one we see around us.

A variation on this idea dates back, in fact, to pre-quantum-theory days. The pioneering thermodynamicist Boltzmann was puzzled by the failure of the universe to be in the most statistically probable state—that of thermodynamic equilibrium. He suggested, in the last century, that perhaps the universe *is* usually in equilibrium, but that occasionally it undergoes vast fluctuations. Once again, the implication is that we see an extraordinary universe around us simply because just such a universe is needed for life to evolve, and hence for us to be here to see it.

Boltzmann's 'fluctuations' were proposed before the big bang origin of the universe was known. But they have a plausible quantum mechanical equivalent. According to this recent idea, the big bang came about as a result of a 'quantum fluctuation' of the universal vacuum.

It is quite possible for, say, an electron and a positron to be created spontaneously out of a vacuum. This is something which has

Above: A nearby spiral galaxy M83 in the constellation Hydra

been known for years, the essential condition for such an event being that certain quantities, such as energy, must be *conserved*. In other words, the total (or net) energy after the event must be the same (i.e. conserved) as it was before the event. The electron and positron, if you like, 'cancel each other out'; their *net* energy is zero. It might seem a little ambitious to stretch this idea to cover the whole universe, but that is what Dr Edward Tryon has done. There is no reason, he says, why the universe should not consist of equal amounts of matter and antimatter, and that takes care of most of the things which need to be conserved in a 'vacuum fluctuation' permissible under the laws of quantum mechanics. The most obvious conservation problem is posed by the positive energy locked up in all that mass—but it seems that this can be 'cancelled out' by the negative energy of the self-gravitation of the universe.

So Tryon calculates that the net energy of the universe may be zero; the same goes for the net electrical charge, and all other conserved properties. In that case, there is no reason why the whole lot could not have appeared spontaneously out of the vacuum some ten thousand million years ago in a big bang, and evolved to its present state. This theory is hardly complete, in a fully satisfying mathematical sense. But in an article in *Nature* in 1973 (Vol 246, p. 396), Tryon pointed out some of its philosophical advantages.

He emphasises that although the scale of the fluctuation is rather large, there is nothing in the laws of physics which makes it impossible. 'Vacuum fluctuations on the scale of our universe are probably quite rare', but since life can only evolve in such a large, long-lived universe, any observers such as ourselves must live in such a universe. 'Our universe is simply one of those things that happen from time to

Left: The business end of the
120-inch telescope at Lick
Observatory showing some of
the sensitive electronic
equipment used in modern
observational astronomy

time.' Of course, such a fluctuation must also have an end as well as a
beginning, presumably in the form of collapse to an ultimate (as far
as this universe is concerned) singularity. That in turn seems, once
again, to suggest a closed universe, full of invisible missing mass.
But, now that black holes are so popular, few astronomers would be
discomfited if the theory brings enough of them in to close the
universe.

This kind of mathematical playing around with the whole universe
is, however, far removed from what most of us mean by astronomy.
It is a little disquieting to ponder for long on just how insignificant
we may be—if the whole universe is just 'one of those things' and
we inhabit an insignificant planet in a remote corner of an obscure
galaxy, our own place seems doubly insignificant. The remarkable
thing is that the inhabitants of that unimportant planet should be
able to probe the mysteries of the universe and have the capacity to
formulate such theories.

But perhaps we have seen enough for the moment of these ultimate
probings of human intellect. There is certainly plenty of activity, as
far as our inquisitive intellect is concerned, in our own astronomical
backyard, where space-probes and other developments have put
studies of the Solar System at the forefront of the new astronomy.

15 Spacecraft & Planets

Now that it is possible to send spacecraft on voyages of exploration to the planets, the study of the Solar System has gained a place of importance in the new astronomy which rivals the role it played in the development of science just after the invention of the telescope. Apart from the exploration of a small part of the Moon, which has been rather comprehensively dealt with elsewhere, unmanned probes have caught the public imagination with a series of epic voyages to the inner planets of the Solar System in the past few years. Both Soviet and American probes have contributed to this great advance of Solar System astronomy. But since far more is known about the American spacecraft, and more is known about NASA's future plans than about those of the Soviet space agency, I shall concentrate mainly on looking at how three or four series of US spacecraft have contributed to the renaissance in planetary science.

Going outwards from the Sun, the first half dozen planets are Mercury, Venus, Earth, Mars, Jupiter and Saturn. Unmanned spacecraft have now investigated all of these planets except Saturn, and Pioneer 11 is now on its lonely way to rendezvous with the ringed planet sometime in 1979.

But for various reasons the exploration of the Solar System has not proceeded in an orderly manner, starting with Mercury and working outwards. The first planet studied from space was, of course, the Earth. These studies have provided a lot of new information about our planet, ranging from accurate measurements of its shape (which turns out to be rather like a lumpy pear) to monitoring of weather patterns, which has opened the way for a new understanding of climate. The impact of those first photographs from space, which showed the Earth as just another planet, was very great, and made many people realise for the first time how important it is for us to maintain the 'life support systems' of 'spaceship Earth'. And by relating studies of the Earth to the new studies of other planets, it is possible to see much more clearly how a planet like ours 'works'.

Leaving out our sister planet, the Moon, the next planet to receive a lot of attention from the spaceprobes was Mars. The red planet has long held a fascination for astronomers; it is one of our nearest neighbours, and all things considered it was (and still is) a prime candidate for investigation. During the late 1960s and early 1970s a flurry of Soviet and American probes have been sent to Mars. On more than one occasion no less than four vehicles have made almost simultaneous assaults on its secrets; but the one probe that revolutionised astronomers' thinking about Mars was NASA's Mariner 9.

Mariner 9 sent back a stream of pictures of Mars, eventually mapping almost the entire planet, and producing 'close-up' large scale pictures of many particularly interesting features. These show that Mars is very much an active planet, and that although it is covered in dust and scarred by meteorite impacts, it has volcanoes and even what seem to be dried up river beds, delta systems, and traces of erosion of the kind caused on Earth by glaciers.

This more than makes up for any disappointment felt when astronomers discovered, some time ago, that the 'canals' of Mars are merely surface features seen through the shifting dust clouds, and are not artificial. Volcanic activity, on however small a scale, hints at underlying geophysical processes rather like those of the Earth. Indeed, one planetary scientist has gone so far as to suggest that Mars has undergone a form of 'continental drift' like the 'tectonic' processes which mould the face of the Earth—but that does rather stretch the analogy between the two planets.

But the real excitement has come from the evidence that water might have existed on Mars in a liquid state, not too long ago. Water, of course, is essential to life as we know it, and if it flows on Mars then there seems a fair chance that some form of plant life might exist on Mars too. The snag is that, according to the best measurements of the pressure and temperature of the atmosphere of Mars, liquid water simply can't flow on its surface today—if any were released from underground (if there is any underground to be released), then it would simply evaporate. That makes a pretty puzzle—but one to which the American planetary scientist Carl Sagan has an equally pretty solution.

The sinuous channels which look so much like dried-up river beds are concentrated in the equatorial regions of Mars, and Sagan points out that water could flow in those regions if the planet could somehow be warmed up slightly. If the heat absorbed at the poles of Mars increased by only 15%, carbon dioxide would be released into the atmosphere as the solid CO_2 of the Martian ice caps (really 'dry ice' caps) evaporated. That would increase the density of the atmosphere —and the 'ice caps' would not form again because the denser atmosphere would carry heat from the equator to the poles.

In other words, according to Sagan, Mars can exist in either one of two stable 'climatic states'. The present conditions he likens to an ice age on Earth; but the warmer conditions would be just as stable, once the changeover had been made, and under those conditions water could flow in the equatorial regions of Mars, because of the increased density of the atmosphere.

What could cause such a change? One idea is that the Sun's heat might vary from time to time by a small amount. This could produce the switchover of the Martian climate—and it could also explain ice ages on Earth. Another theory is that the wobble of Mars as it spins round the Sun like a giant top will sometimes change its tilt, so that one pole points more directly towards the Sun, and so gets more heat. According to some theories, that wobble could result in regular

Right: Surface features of Mars photographed from Mariner 9 showing features which suggest erosion by water or glaciers. Below: Phobos, one of the moons of Mars photographed by Mariner 9. The irregular shape of Phobos suggests that it is a lump of cosmic rubble captured by Mars from the asteroid belt

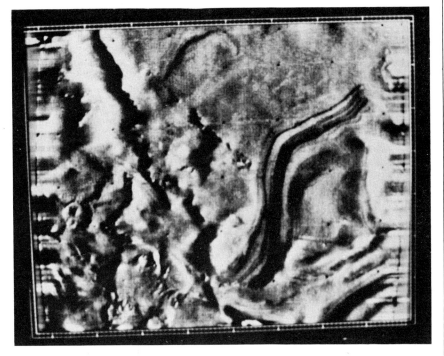

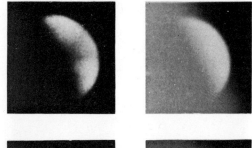
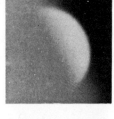
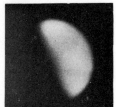
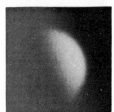

Left: Earth-based telescopes give only a vague impression of the surface detail of even the nearest planets such as Venus (left) photographed at ultra-violet and infra-red frequencies. Space probes such as Mariner 9 provide far more information. Below: A Martian canyon wider and deeper than the Grand Canyon in Arizona, photographed by Mariner 9, is shown here with the changing relief indicated

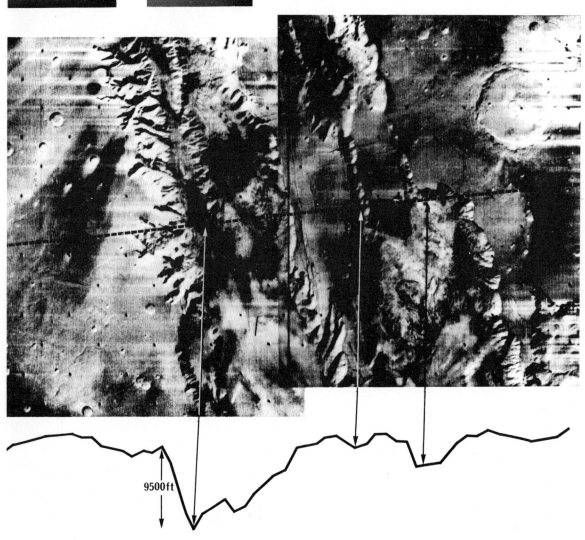

9500ft

Right: A close-up of part of the Martian 'Grand Canyon'.
Below: A complete map of Mars based on Mariner 9 photographs. Middle latitudes are shown in the bottom map, with north polar regions above and to the left, and south polar regions above and to the right

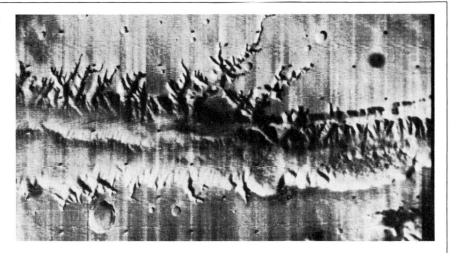

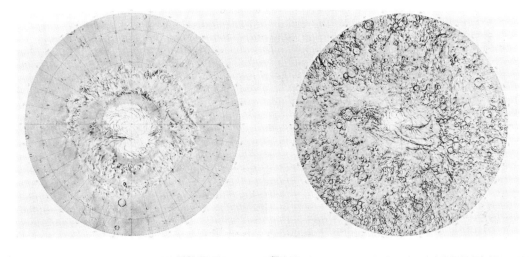

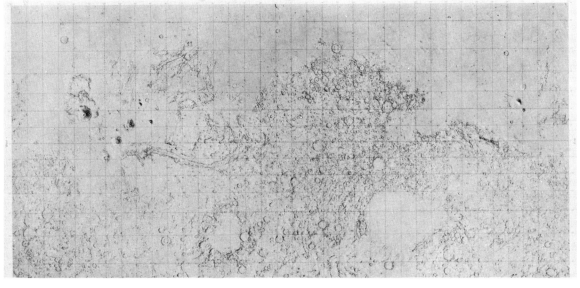

periods of warm and cold climate, lasting for millions of years.

But the most attractive idea of all is even more straightforward. The same Mariner vehicle which showed the 'river beds' on Mars also sent back detailed pictures of a great dust storm. If a lot of dust from such a giant storm was dumped on the polar caps, they would absorb more heat, because dust is less reflective than CO_2 'ice'. And that could switch conditions out of the present ice age—although then you have still got the problem of explaining how the ice age developed in the first place.

Sagan's best speculations about Mars, however, depend on there being some more or less regular switching between the two climatic states, even though each state may remain stable for millions of years at a time. During the 'warm spells' some kind of life could evolve, and produce spores which might survive the long, intervening ice ages. Spores can survive for very long times on Earth, where there has been no evolutionary pressure to develop this survival ability, so why not on Mars? In the normal run of things, we could expect that sometime in the distant future the Martian climate would make its regular switch; then water now frozen below the surface might flow once again, bringing about a remarkable 'spring' as spores are triggered into life.

We may not have to wait too long to find out if the theory works. In the summer of 1976 NASA plans to land Viking spacecraft on Mars. Unlike Soviet spacecraft which were the first to achieve the remarkable feat of a soft landing on another planet, the Vikings will carry water with them, and two of the experiments planned for the robot spacecraft involve adding water to samples of Martian soil. If any of those samples contains one of the spores Sagan has speculated about, the results could be spectacular, to say the least, and those results may already have made headlines by the time you read this.

It might seem logical that, having started with our near neighbour Mars, the next development of spaceprobe-assisted planetary science would have involved Venus, our nearest neighbour on the other side. But Venus has proved something of a disappointment. Soviet equipment parachuted down through the planet's atmosphere, and American 'flyby' spacecraft, have all painted a grim picture of Venus.

Once beloved of science fiction writers as potentially the most Earth-like of all the other planets of the Solar System, Venus is now revealed as a planet completely covered in clouds (which contain sulphuric acid droplets) and with a hot, dense atmosphere quite uncongenial for terrestrial life. The latest pictures of Venus, obtained from Mariner 10 (of which more later) show interesting patterns in the clouds, and there are definite layers, or strata, in the cloud cover. Certainly Venus is going to keep the planetologists and space scientists busy in the years ahead—but as yet it has not turned up with the kind of exciting developments that have made recent investigations of Mercury, Mars and Jupiter so dramatic.

In fact, at the end of 1973 it was a mission to Jupiter which made the headlines (just before an equally successful unmanned probe to

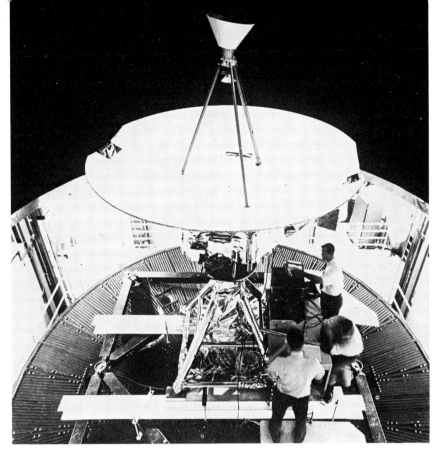

Mercury). The spacecraft was Pioneer 10, which has now flown past Jupiter and is proceeding out into interstellar space—the first man-made object to leave the Solar System. Not least among the remarkable achievements of Pioneer 10 is that it continues to work perfectly even out beyond the orbit of Jupiter, sending back valuable information about interplanetary space.

Jupiter's orbit is, on average, 778 million kilometres from the Sun, and the distance from Earth to the Pioneer 10 spacecraft when it passed Jupiter was so great that radio signals, travelling at the speed of light, took almost three quarters of an hour to get back to Earth. Yet the pictures sent back surpassed any obtained by telescopes on Earth—and the scientific experiments provided even more exciting information about the largest planet in the Solar System.

Perhaps the most important discovery is that Jupiter dominates interstellar space for a vast region around itself. Jupiter rotates with a period of just under 10 hours and produces streams and bursts of cosmic ray particles which show this characteristic 10-hour variation. These particles were detected by Pioneer 10, six months before it reached Jupiter—that is, 100 million miles away. Within an enormous volume of space 100 or 200 million miles across, it is Jupiter, rather than the Sun, which dominates interplanetary space.

This provides astronomers with a new way of studying what goes on in interplanetary space, although as yet they have not had time to find anything out from the discovery of Jupiter's sphere of influence.

Indeed, that is characteristic of all of the Pioneer 10 information. As the name of the spacecraft suggests, it was intended to pave the way for others, proving that such flights are possible and that valuable scientific information can be sent back from so far away. The actual information sent by Pioneer 10 itself is almost to be regarded as a bonus—although it is that 'bonus' which has mapped the radiation fields around Jupiter and provided new information about the planet's radio emission, as well as sending spectacular pictures back to Earth.

Pioneer 10's discoveries have already allowed NASA scientists to make best use of another spaceprobe, Pioneer 11, a sister spacecraft to Pioneer 10, which was launched towards Jupiter a year after its predecessor. In the words of a NASA report, the route to Jupiter and beyond was proved 'dangerous but survivable' by Pioneer 10. Pioneer 11 was intended as a backup to Pioneer 10 in case of mechanical failure. But with Pioneer 10 successful, and the route proved, NASA scientists were keen to press on further with Pioneer 11.

So Pioneer 11 was retargeted to swing past Jupiter and use the gravity of that giant planet to pull it onto a new course, towards Saturn. To use the 'gravitational slingshot' effect properly, so that Jupiter swung Pioneer 11 onto the correct course, the spacecraft passed closer to Jupiter than its predecessor. That was only possible because Pioneer 10 found a 'hole' in the damaging radiation belts through which Pioneer 11 could fly—so the new mission really does build upon the success of its predecessor. And Pioneer 11 has produced one of the most dramatic scientific discoveries of the entire space programme.

Jupiter has been known as an emitter of natural radio noise for twenty years. The intensity of the radio emission indicated that the planet must have a strong magnetic field, because radio noise can be produced by the movement of charged particles, such as electrons, in such a field. As a first guess, astronomers assumed that the Jovian magnetic field might be rather like that of the Earth—a simple dipole field, with one north pole and one south pole. That simple picture was not completely satisfactory, but even when Pioneer 10 passed within two Jupiter diameters of the planet's surface in December 1973 it found no evidence to change the picture drastically.

There was some risk that Pioneer 11's closer approach, through the magnetosphere of Jupiter, might damage the space probe, but even complete destruction of the instruments would have provided useful information. As it happens, Pioneer 11 continued to function normally in spite of its close swing past Jupiter—and the information obtained fully justified the risk of damage.

According to these measurements from close to Jupiter, made in December 1974, the magnetic field of that planet cannot be explained in terms of a simple dipole. The next most complicated field would be a quadrupole, which can be imagined as a combination of two north poles and two south poles. But, to explain the Pioneer 11 observations, Dr M. H. Acuna and Dr N. F. Ness, of the NASA-Goddard Space Flight Center, had to go one step further still.

Right: Apollo 11 on the Moon
with astronaut E E Aldrin
deploying scientific experiments

The two NASA scientists invoked the presence of an 'octupole' field to account for the variations detected by the instruments on-board Pioneer 11. Although such a field can be imagined as the field that would be produced from a certain superposition of four north and four south poles, this does not really mean that there are eight 'poles'—or four dipoles—on Jupiter. The field is undoubtedly generated by electromagnetic processes, with an interrelated flow of currents producing a field more complicated than the dipole field produced by a simple coil or a bar magnet.

One of the exciting features of this discovery is that it could only have been made by actually sending the measuring instruments to Jupiter. If there are still any scientists around who doubt the value of using spaceprobes of this kind and would prefer to rely on ground-based observations, they must surely change their tune now. As Drs Acuna and Ness say, this 'octupole model of the main magnetic field of Jupiter will permit reconciliation of the previous twenty years of radio astronomy data with the real magnetic field', and could lead to 'an understanding of the basic source mechanisms and processes leading to the ... emission'.

The actual path followed by Pioneer 11 is almost as interesting as the mission itself. Swinging past Jupiter late in 1974, the space-craft was diverted back across the Solar System, and will eventually meet Saturn on the opposite side of the Sun from the Jupiter encounter. But that long voyage will take almost five years, with the

Saturn encounter not expected until October 1979. Saturn is 9·5 times as far from the Sun as the Earth is, and signals from Saturn will take 1·5 hours to reach Earth. Clearly, Pioneer 11, or however much of the satellite as survives the next five years in space, will not be able to report in great detail about the ringed planet. But as long as the spacecraft is working at all, it will be useful simply to fly it through the rings of Saturn to see if it hits anything solid, or whether the rings are insubstantial collections of ice and dust grains.

That would be enough to pioneer the way for spaceprobes of the 1980s. But if we (and NASA) are very lucky, it is just possible that by the end of 1979 we will have, courtesy of Pioneer 11, the first close-up pictures of the most beautiful object in the Solar System— Saturn, the ringed planet.

But that is still a long way off, and meanwhile we should consider one more planet which has already been investigated—only by one spaceprobe, it's true, but that one vehicle has completely changed astronomers' views. We are talking about Mercury, the closest planet to the Sun.

Because Mercury is so close to the Sun, it is very difficult to observe from Earth, and little was known about it until 1974. Then the spacecraft Mariner 10 was sent on a close flyby of Mercury.

This Mariner 10 is the same spacecraft which flew past Venus a few months earlier. Using a 'gravitational slingshot' effect, as Pioneer 11 did later in the year with Jupiter, Mariner 10 was swung on towards Mercury by the gravitational pull of Venus.

Once again, the NASA vehicle sent back superb pictures of its target. We have become so used to seeing such very good pictures that it is easy to take them for granted. But apart from the other scientific information sent back by these various spacecraft, it really is a remarkable technical achievement to get pictures of such quality; in the case of Mercury, the Mariner 10 pictures show more detail of that planet than the Lunar orbiter spacecraft pictures of less than ten years before showed of the Moon.

That comparison is particularly appropriate—the Mariner 10 pictures show that Mercury is cratered in a way superficially very like the lunar surface. This emphasises the broad similarities between the inner planets of the Solar System. Planetologists are now sure that Venus too will be found to have received its fair share of meteorite impacts—although on Venus, as on Earth, the craters will have been smoothed out by erosion.

On the Moon, of course, there is no wind (since there is no atmosphere) and so no erosion. On Mars, the thin atmosphere allows some erosion, and this is shown by Mariner 9 pictures. Mercury has very little atmosphere indeed, which is why its craters are so clearly defined. But it does have some atmosphere, much to the surprise of most astronomers.

It was always accepted by the majority of astronomers that Mercury could have virtually no atmosphere at all, because the heat of the Sun would drive any gases off from such a small planet so close to the

Right: The cratered surface of Mercury shown here in Mariner 10 photographs from a distance of about 400,000 km (left) and 20,700 km (right) closely resembles our own Moon

Sun. Clearly they were wrong, since Mariner 10 reports a small but measurable trace of an atmosphere. And the reason why they were wrong is easy to see—with hindsight!

Mercury has a small magnetic field, about 1% of that of the Earth; but that is enough to play a part in trapping charged particles from the solar wind, which is, of course, relatively strong at the orbit of Mercury. In addition, radioactive processes going on inside the rocks of Mercury will release traces of gases such as helium, or alpha particles (from the radioactive decay of uranium) and argon (from the radioactive decay of potassium). Together, these effects could produce enough atmosphere to explain the Mariner 10 observations.

That provided a salutary lesson to those planetologists who still did not believe in the value of spacecraft. But to balance this, Mariner 10 also provided evidence in support of one accepted theory of planet formation.

It turns out that Mercury is quite massive for its size, with a density (8 gm/cm^3) eight times that of water. This can only mean that it is made up largely of iron or similar elements—perhaps as much as two thirds of the planet must be iron-group elements; and that, of course, helps to explain its magnetic field. The importance of this to planetary theory is that, according to the best theory, the planets and Sun collapsed together out of a primeval cloud of dust and gas. As the Sun 'switched on' and grew hot, it would have driven off any light gases and elements from the innermost planets, leaving fairly small, dense bodies. Further out, the lighter elements could form gas giant planets, like Jupiter and Saturn; but the inner planets could only, according to the theory, be solid and rocky like the Earth. With the latest Mercury measurements in from Mariner 10, the gradation in densities of the planets measuring outward from the Sun is quite clear. As theory predicts, Mercury is indeed the densest planet, since almost everything except iron was boiled out of the inner Solar System when Mercury was forming.

Like studies of Jupiter and Mars from spacecraft, however, the study of Mercury is only just beginning. At last, astronomers are able to piece together a coherent picture of the Solar System and how the Earth fits into it. At the same time, the equipment now becoming available is opening up the way for what will probably be the golden age of astronomy as a whole, leading to a real understanding of how the Solar System 'fits' into our galaxy, and our galaxy into the whole universe. So far, we have seen the merest hint of how the new astronomy is developing. What is likely to happen in the near future, as astronomers strive to improve their understanding of our changing universe?

16 The Future

As we have seen, man's view of our universe has changed over the past three decades because the new astronomy has opened up new 'windows' through which observations can be made. First radio astronomy provided a new means of investigating the universe from the ground, then, with the advent of artificial satellites, it became possible to make observations at literally any part of the electromagnetic spectrum.

At longer wavelengths, this electromagnetic radiation is called radio; as the energy of the radiation increases, and its wavelength decreases, we have microwaves, infrared radiation, and then ordinary visible light; and on the high energy side of visible light we have ultraviolet, gamma and X-rays. All of these can now be observed either from the ground, from balloons, or from space.

Every time a new part of the spectrum has been opened up to observation, new and surprising discoveries have been made—radio galaxies and X-ray stars are obvious examples. It might seem that now there is little left to look forward to, since no such breakthrough into a new part of the spectrum can ever occur again. But that is not the case. Pulsars, for example, were only discovered decades after radio astronomy began, and who can tell what new phenomena might be discovered at other frequencies during the next few years?

In addition, astronomers are now in a position where they have more information flowing in than they can handle. Uhuru has now recorded the existence of a couple of hundred X-ray sources, but very few of these have been examined in detail. Just because all regions of the spectrum have been opened up doesn't mean that they have all been explored—it's better, perhaps, to compare the present situation in astronomy with the first glimpse of a new continent. The outlines of part of the shore can be seen and even mapped in a general way, but we have still to set foot on the continent and explore the interior.

So the next ten or twenty years will probably be remembered as the golden age of astronomy. Whatever your interest, there is something relevant to it in the new astronomy—and no one is likely to object to another pioneer helping to explore the 'new continent'. Some of the likely developments can now be guessed at—but there will always be the unexpected as well.

Perhaps the path ahead is more clearly marked in X-ray astronomy than in some other areas. The first priority, of course, is simply to find out more about the hundreds of X-ray sources now known. And that means locating them with pinpoint accuracy, to see if they can be identified with particular stars.

We have already seen, in Chapter 10, how the satellite Copernicus can use its specialised directional telescopes to pin down the positions of X-ray sources fairly accurately. But even better accuracy is needed; and with the experience of radio astronomers to draw on, X-ray astronomers are now developing the lunar occultation technique mentioned in Chapter 3. It is rather harder, of course, to work out all the directions involved when the measurements are being made by a satellite in orbit about the Earth, and a lot of computation is needed to work out the position of an X-ray source from measurements of the time when it is 'eclipsed' by the Moon.

But the technique does work, as was proved in 1973 by a team of British astronomers using Copernicus. Indeed, the technique worked so well that they located the position of the source GX5 + 1 to within an 'error box' 6 × 20 seconds of arc across, compared with a previous best estimate of ±1·5 minutes of arc.

The snag with this, however, is that the Moon covers such a tiny part of the sky, and only ever occults a few sources around the equator. That is why lunar occultation has only been used on a few radio sources. But with an X-ray satellite, the observers can have more flexibility, as ESRO (now part of the European Space Agency, ESA) realised. If a satellite is launched into an eccentric orbit which takes it well out of the plane in which the Moon orbits the Earth, then it will appear from the satellite as if the Moon wanders about over almost all of the sky.

Just such a satellite—EXOSAT—is now planned. When it is launched, it will be possible to apply the lunar occultation technique to almost every known X-ray source, over a period of time, as the satellite wanders around in its peculiar orbit. And that is bound to bring X-ray astronomy forward by another large stride.

Radio astronomy, too, can benefit from observations above the Earth's atmosphere and away from man-made radio interference. Again, several possibilities are being considered; but one which seems likely to prove particularly useful is a radio astronomy satellite which will orbit not the Earth but the Moon. That should be able to detect faint sources of radio emission which are 'invisible' against the radio noise on Earth.

Observations in the infrared are already providing significant information about relatively cool clouds of gas and dust, which might very well be young stars in the making. These objects, it seems, have not yet warmed up enough to emit much at higher energies; one hope is that in the not too distant future astronomers who are studying these infrared clouds might actually see a new star 'switch on' as nuclear reactions start up in its interior.

On the other side of the visible spectrum from the infrared, gamma rays are just beginning to provide useful information to astronomers. Here, there has been, compared with the excitement of X-ray astronomy, something of a long, hard struggle. Gamma rays are not too difficult to detect, using modern counters, but since they can

Opposite left: Two photographs of the spiral galaxy NGC2442, obtained with the 4-metre Anglo-Australian Telescope. The top picture is a long exposure showing clearly the faint outer regions; the bottom picture, a shorter exposure, reveals details of the spiral structure. Opposite right: Another spiral galaxy, NGC1566 (top) and the giant globular cluster 47 Tucanae (bottom), a collection of at least 10 million stars forming a gravitationally bound system within our own galaxy. Astronomers often print photographs in negative form, making it easier to pick out faint objects

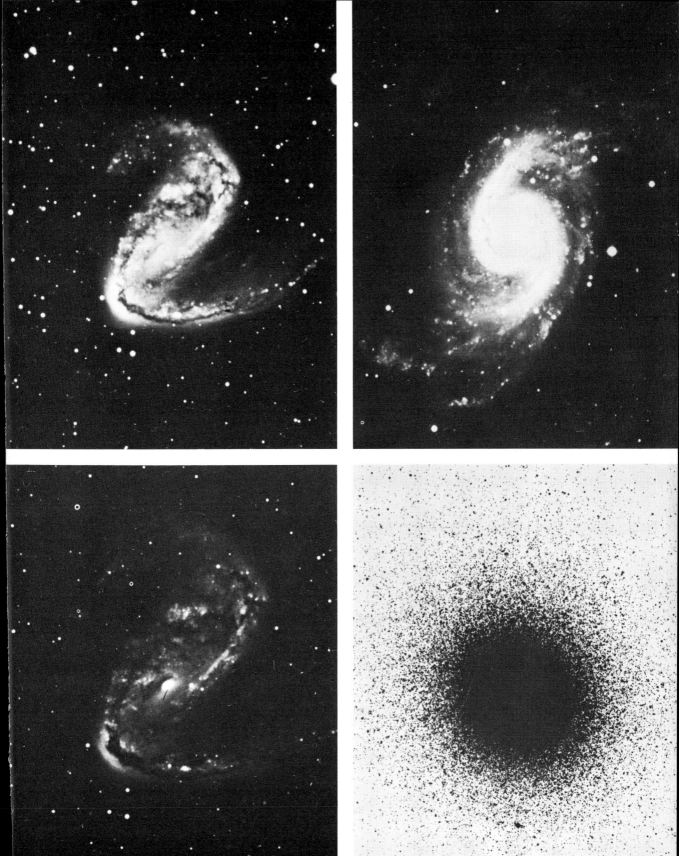

penetrate the sides of most satellites it's not always easy to tell where they are coming from. Indeed, experiments on some satellites have been confused because cosmic rays strike the satellite, or other experiments carried on it, and react to produce a shower of secondary cosmic rays and gamma rays, which are duly recorded on the gamma ray detectors! The only really good way of pinning down the direction of gamma ray bursts is by a sort of triangulation, measuring the time of arrival of the burst at two or three satellites, and thence working out where it came from.

But at the beginning of experimentation in gamma ray astronomy it wasn't even clear that anything interesting was being detected at all. The breakthrough came when the Vela satellites were launched, to provide a system for monitoring nuclear tests on Earth. These satellites include gamma ray detectors, since nuclear bombs produce lots of gamma rays. But although they found a good number of gamma ray bursts, the rays were not coming from Earth.

One of the most interesting things about these bursts is that they last for a very short time—a few thousandths of a second. According to one theory of supernova explosions, that is just what should be expected if the gamma rays come from the explosion of whole stars— the same sort of thing as the nuclear explosions the equipment was designed to monitor, but on a rather larger scale! That would mean that the gamma rays being detected come from supernova explosions in distant galaxies, since such events only occur once or twice in a galaxy in a hundred years.

So far, only about thirty gamma ray bursts have been detected in four years, roughly one for each group of theoretical gamma ray astronomers. It's quite possible, on the basis of this slender evidence, that the bursts don't come from distant supernovae at all, but originate quite close to home. So this is one area of research where very rapid progress is likely to be made as more bursts are studied over the next few years. The great breakthrough will come, of course, when someone is able to locate the position of a gamma ray source so accurately that it can be identified with a star or galaxy, or perhaps with a pulsar.

Within the Solar System, spaceprobes to make soft landings on Mars (due for launch in the mid-1970s) and Venus,* and further probes to Jupiter, Saturn and beyond will continue to keep planetary scientists busy. But it is not only through the use of spacecraft that the new astronomy will continue to develop.

By modern standards it might seem that ground-based radio astronomy is almost old hat. But of course the bulk of radio work is still done from the ground, and every time anything interesting is discovered by other means, radio astronomers are asked to investigate it with their telescopes. This puts enormous pressure on the time available, and bigger and better radio telescopes are still needed and

* Two Soviet probes reached Venus and sent back pictures and scientific data from the surface in 1975.

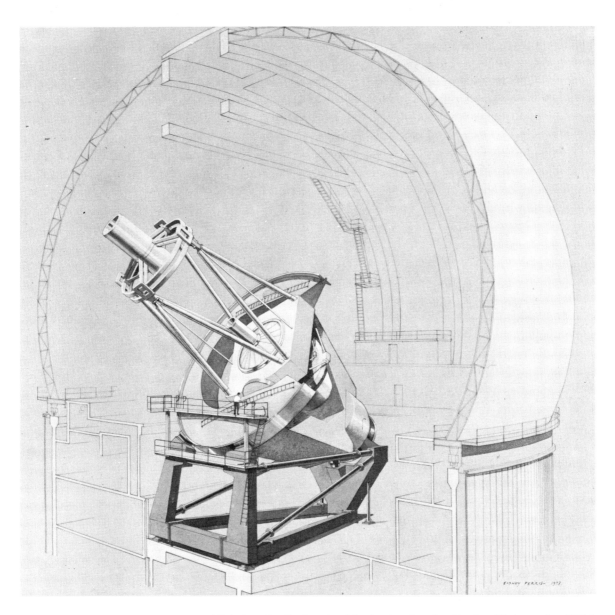

Above: An engineer's drawing
of the Anglo-Australian
Telescope

are still being built. In the Netherlands, the Soviet Union, the USA and Australia, in particular, new instruments are always being developed. In Britain, the Cambridge 5 km telescope is the latest development, and although plans for another instrument, the Mark V 'Jodrell Bank' telescope, have had to be shelved because of the present economic situation, it seems very likely that the existing instruments at Jodrell Bank will be completely overhauled to keep them at the forefront of future developments. Such telescopes are expensive, of course, involving many thousands of tons of steel in their construction. But the Science Research Council, which has to foot the bill in the case of the Jodrell Bank work, can take comfort from the fact that, as steel has become scarcer, the Mark I itself is now worth more as scrap than it cost to build! But of course, there are no plans to scrap this valuable instrument, which continues to be one of the major tools of radio astronomy.

Indeed, it is astonishing how few astronomical tools ever seem to be scrapped. They are always being improved in one way or another, doing work that their designers never envisaged. And it would be very wrong to think that optical astronomy has reached the end of its development.

In optical astronomy there is some debate about whether the limited money available should be spent on one or two big telescopes, like the 200-inch, or on a series of smaller instruments. The advantage of having several smaller telescopes is, of course, that you can look at more things at once, even though a large telescope might reveal more about each object on which it is focused. With so many stars and galaxies in the sky, that is a powerful argument. And any development which speeds up the use of telescopes is eagerly taken up by the astronomical community.

One such development has been made very recently. Virtually all astronomical observations involve using instruments or photographic equipment—attached to a telescope, of course, since the human eye is quite inadequate for detailed studies. The recent breakthrough, made by a team from the Scottish Royal Observatory in Edinburgh, involves the speeding-up of astronomical photography.

Astronomers measure brightness on a magnitude scale in which 'first' magnitude was defined historically as the brightness of a candle flame at a distance of 1300 feet. That figure has now been replaced by calibration in terms of bright stars—but it indicates roughly what a first magnitude star looks like to the unaided eye.

Just to make things confusing, each numerical step upwards in magnitude (1, 2, 3 etc.) corresponds to a *decrease* in brightness by a factor of 2·5. So an increase of five magnitudes means a decrease in brightness by a factor of 100. (The Sun, incidentally, is magnitude −27, which means that it is about $2·5^{28}$ times *brighter* than a candle at a distance of 1300 feet.) The human eye can, under ideal conditions, see stars as faint as magnitude 6; but photography using astronomical telescopes with long time exposure can produce images of stars as faint as magnitude 23—more than a million times fainter than the faintest stars visible to the eye.

The catch in this, of course, is the long hours needed to get the photographs of the faintest objects. It's not only tedious and inconvenient, waiting for cloudless conditions and building telescopes in Chile where the air is clearer; but if each photograph can be made a little more quickly then the telescope is available for a little more time to look at other things. So astronomers are always trying to find ways of making the photographic emulsions they use more sensitive.

Various techniques, including baking the films and putting them in vacuum chambers, are used to achieve this. They work because they get rid of impurities in the emulsion, like water and oxygen. But a

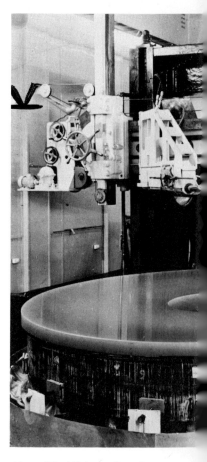

Above: The 148-inch mirror for the Anglo-Australian telescope being prepared at the Grubb Parsons' Works in Newcastle upon Tyne

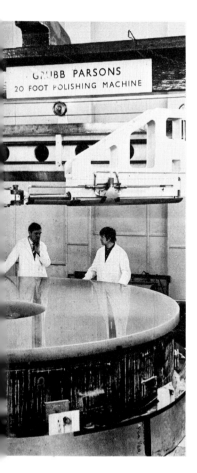

plate which has been baked and exposed to vacuum is not always left in the best condition. The emulsion tends to shrink, and it can be fogged by the treatment. Within the past couple of years, however, the Edinburgh team have found a way to remove oxygen and water without undesirable side effects.

This new technique involves 'soaking' the plates in a sealed tank containing nitrogen. Water and oxygen escape from the emulsion, the nitrogen is flushed out, and the process repeated for a month. By then, few impurities are left.

So now astronomers have a new and better way of increasing the sensitivity of their plates. As a result, new surveys of the sky are recording fainter objects than those in previous surveys. The classic survey, or photographic map, of the northern hemisphere sky was made at Mount Palomar some twenty years ago. It shows everything down to a magnitude of 21. But today, a new survey is being made of the southern sky, jointly by British astronomers in Australia and European astronomers in Chile. For this survey, the new nitrogen-soaking treatment is being applied to the plates, and together with other improvements since the 1950s, this means that the new survey is complete down to magnitude 23. Objects only one sixth as bright as the faintest objects on the Palomar survey will appear on the new plates.

This is of the greatest importance in trying to identify distant radio galaxies and quasars. Indeed, it may well be time to carry out another survey of the northern hemisphere sky, using the new techniques and emulsions. And so, like the painting of the Forth Bridge, the sky survey is probably a never-ending task.

Such ingenuity is bringing out new aspects even of the oldest branch of astronomy, and there can be little doubt that comparable efforts applied to what are now the new fields of astronomy will continue to result in exciting new developments for many years to come. The remarkable progress of the past twenty years is just the beginning of the new astronomy. With all the new ground-based and space-based systems available and beginning to provide information, we have certainly not reached the end of modern astronomy, or even the beginning of the end. Better, perhaps, to say that now that the entire electromagnetic spectrum is available to observation, together with cosmic rays, we have in fact reached the end of the beginning of the new astronomy.

Appendix–Ice Ages, Man & Solar Neutrinos

Although it's convenient to divide the study of our changing universe up into sections, much of the appeal of astronomy—and its value—lies in the fact that there are no really rigid compartments. Astronomers who might seem to be specialists in X-ray or radio studies, and theorists, are all eager to find time to use conventional optical telescopes to find out more information about the objects they are studying. The observers themselves, once thought of to some extent as data gatherers, are now expected to develop their own theories to explain the observations; the fact that the present Director of Britain's Royal Greenwich Observatory made his reputation as a radio astronomer and theorist at Jodrell Bank highlights this non-specialist approach to modern astronomy. In many ways, the most valuable asset for an astronomer is to have the imagination needed to fit what seem to be unrelated bits of information about different aspects of the universe together in a broad picture. Nowhere is this approach shown better than in recent work by Professor W. H. McCrea, of Sussex University, who has developed a theory tying together Ice Ages, the origin of Man and the formation of comets and the Solar System in terms of the overall structure of our galaxy. In fact, even that is not the end of the story, since I have been able to extend McCrea's theory to provide a possible explanation of the solar neutrino problem (see Chapter 8)! That, of course, is the way most theorists work—building a small extension onto the towering edifice of some great man's work. First, take a look at the broad picture.

Anyone who attempts to predict Ice Ages needs a bold imagination, but most such bold attempts in recent years have proved rather gloomy, suggesting that a new Ice Age may be imminent (see my book *Forecasts, Famines and Freezes*, Wildwood, London and Walker, New York, 1976). Part of the appeal of McCrea's bold imaginings is that, as well as their breathtaking astronomical sweep, they produce both the longest and the happiest long range forecast ever: no more Ice Ages for at least 100 million years. The reason why McCrea's views differ from those of most other crystal ball gazers is that he has looked not just at the Earth as a whole, but at the place of the Earth in our galaxy. On that scale, the changes studied by most climatologists are minor details compared with astronomical effects, and the astronomical effects may well dominate in the long term, although of course they play little part in deciding what the weather is like from day to day or year to year.

In very round terms, according to McCrea, the geological evidence suggests that periods of ice cover recur on Earth roughly at intervals of 250 million years. McCrea calls these periods Ice Epochs, to distinguish them from Ice Ages; each Ice Epoch may last for several million years, and will cover several Ice Ages, and several rather warmer periods, interglacials, when conditions will be rather like those on Earth today. Such minor fluctuations, on timescales of a few million years or less, are the fine details which are of little concern to McCrea's astronomical theory, but provide the basis of conventional palaeoclimatic studies.

Now, this period between Ice Epochs is just half the time it takes our Solar System to move once around the galaxy, orbiting the galactic centre in much the same way that the Earth orbits the Sun. This 'coincidence' was noted more than fifty years ago—but could it be more than a coincidence? If it is, then the effect must presumably be related to the spiral structure of our galaxy, since the two spiral arms will each be crossed once by the Solar System in each orbit. McCrea's bold step forward was to provide an explanation of how the two passages across spiral arms every 500 million years could cause Ice Epochs.

So-called spiral arms are a feature of many galaxies (see, for example, M51, p. 102 and on the title page). The bright pattern is caused by the presence of many hot, young stars along the spiral path, but the true, permanent features which produce the pattern are the dark lanes of dust and gas which can be clearly seen at the edges of the bright arms. According to the best current theories, the concentration of dark gas and dust marks the permanent spiral feature, a standing shock wave through which stars must pass as they orbit the galaxy. How the shock originates is another, as yet unanswered, question. But knowing that it does exist, how will it affect the stars—or solar systems—which pass through it?

When the Solar System runs into such a concentration of dust, the first effect is that the Sun will be warmed up a little. This is because dust falling into the Sun loses its gravitational energy in the form of heat (rather like a smaller scale version of the way energy is produced in X-ray stars, see Chapter 10). Paradoxically, a small warming of the Sun can lead to an Ice Epoch on Earth; what seems likely to happen is that the warmth at first increases evaporation from the oceans, producing more clouds, and that the clouds reflect away not just the extra heat but a lot more as well. Once ice sheets form under the clouds, they will continue to reflect a great deal of the Sun's heat away even if the clouds disperse.

Of course, this only works for small changes in solar output. With a big increase in heat, the Earth would get hotter even if clouds were reflecting some of the heat away. But according to McCrea's calculations, the changes produced when the Solar System crosses the lane of dark material in front of a spiral arm would be just right to produce this effect. Even better, the model can explain the present situation on Earth very accurately.

It seems that the Solar System is now on the edge of a spiral arm (the Orion Arm) and has recently (by astronomical standards) passed through the associated lane of dust and gas. According to

McCrea, we may even have passed through the Orion Nebula (see p. 78–9); and we emerged into clear space about 10,000 years ago. When did the last Ice Age end? Just about 10,000 years ago—so if McCrea is correct, that Ice Age was the last of the latest Ice Epoch, and as we continue through the Orion Arm and beyond there will be no more Ice Ages until we reach the next spiral arm. That will be in about 250 million years if you just take half the period of the Earth's orbit around the galaxy; being cautious by nature, McCrea only goes so far as to suggest that things should be all plain sailing for the next 100 million years, which allows a margin for error if astronomers are slightly wrong about the time it takes for us to complete one orbit!

But that's far from being the end of the story. McCrea has also looked into other effects of the Sun's passage through these dust lanes, and he comes up with some even more fundamental speculations. The bright, young stars which mark the spiral arms are believed to have formed when clouds of gas orbiting the galaxy were compressed by the shock wave of the spiral pattern. That explains why there are so many of them just behind the dark lanes; by the time they move far round the galaxy they have become older, cooler and less bright, and that explains the appearance of the bright spiral pattern. So McCrea suggests that our own Solar System formed in just this way, but hundreds of millions of years ago. The fascinating result of his calculations is the implication that as a cloud of gas moves around the galaxy it will become more compressed each time it passes through the shock wave of the spiral pattern. Eventually, when a critical density is reached, the cloud can collapse under its own gravity to form a star; but for intermediate cases, it's likely that clouds will condense out to form planetary sized objects, and for low densities the nuclei of comets can form by the compression of these interstellar clouds.

On that picture, the comets which now form part of the Solar System are a by-product of the recent passage through a compression lane, captured by the Sun en route. It may be that we are lucky to see any comets at all, and that for most of the long history of the Solar System there have been no comets bound to the Sun. Those we have at present will soon be dispersed, or broken up by repeated passage into the inner Solar System and past the Sun.

So our Solar System may be the offspring of the spiral structure of our galaxy. Since the kind of climatic variations explained by McCrea's theory are likely to be of great importance to any life forms emerging or developing on a planet, it looks as if the development of life depends not only on the distance of a planet from its parent star, but on the distance of the star from the centre of its parent galaxy, and the frequency with which it encounters spiral arms. Even for astronomy, a theory which ties together the origin of the Solar System and comets with the emergence of life on Earth and the occurrence of Ice Ages is imagination on the grand scale—but even that is not the end of the story.

When I first heard of McCrea's theory, it occurred to me that the kind of temporary variation of the Sun's radiation which seems to be caused by encounters with dust lanes might be just the effect needed to resolve the solar neutrino 'problem'. As we saw in Chapter 8, the absence of solar neutrinos today might be because the Sun is not at present in the usual state of stellar equilibrium. If something in the recent astronomical past has happened to stir up the solar interior, the nuclear reactions which should produce the neutrinos may be temporarily 'switched off' while the Sun is settling down. So what would be the effect of heating the Sun's surface by gathering up interstellar dust?

In fact, the addition of the dust probably has very little effect on the interior. By expanding or contracting a little—using its gravitational potential as a safety valve—the Sun can accommodate large surface changes without correspondingly large interior changes. At present, parts of the Sun should be very close to 'convective instability', according to our best theories. In other words, the right kind of trigger could induce widespread convection, which is just what we would need to explain the absence of solar neutrinos. But adding dust in line with McCrea's theory seems, if anything to reduce the size of the convective zone in the Sun. It is when the extra heating from the dust is removed that convection increases again back toward its usual value—and if that increase overshoots the 'normal' state by even a small amount, it could be enough to allow the growth of convection in just the way several theorists have proposed to resolve the solar neutrino problem.

It's a bit like taking the lid off a pressure cooker while it is still on the stove—the sudden change will change the equilibrium below and it will take some time to settle down again. Just how long it would take in the case of the Sun is not certain, but it would probably be at least a few million years, and perhaps more than 10 million years, in line with the well known solar timescale, the Kelvin-Helmholtz time. Since we emerged from the dust only a few tens of thousands of years ago, there seems no doubt that the solar pressure cooker will not yet have adjusted to normal conditions, and that no solar neutrinos are being produced today.

It's ironic that some astronomers had already suggested a link between the solar neutrino problem and Ice Ages, but without suggesting that they were the separate by-product of a much bigger phenomenon. As far as the neutrino problem is concerned, it's too early yet to say that my extension of McCrea's theory has provided the final answer that astronomers have been waiting for. But this whole story of Ice Ages, dust and solar neutrinos does show how even a shallow knowledge of astronomy can lead to important new insights, provided that the shallow knowledge is spread broadly enough. Of course, if you want to reach the heights, like Professor McCrea, then you need a knowledge both broad and deep; even then, however, the new astronomy depends as much on a synthesis of wide ranging ideas as on any narrow, specialist study of one topic.

Bibliography and further reading

Radio Astronomy
Hey, J. S. *The Evolution of Radio Astronomy* Elek, 1972
This is probably the best 'popular' history of radio astronomy.
Lovell, B. *The Story of Jodrell Bank* Oxford University Press, 1968
Piper, R. *The Story of Jodrell Bank* Carousel, 1972

Cosmology
Sciama, D. W. *Modern Cosmology* Cambridge University Press, 1971
This is an up-to-date account by an eminent modern cosmologist, for those interested in serious investigation of the subject.

Stars
Tayler, R. *The Stars* Wykeham Science Series, 1970
This is an introductory text for the more serious student of astronomy.

Galaxies
Shapley, H. *Galaxies* Harvard University Press, 1972
This is the best introduction, even though it was originally written in 1943 and last revised more than ten years ago.

Pulsars
Pulsating Stars Macmillan, 1968
Pulsating Stars 2 Macmillan, 1969
These are two books of reprints of scientific papers first published in *Nature* which capture the excitement of pulsar work in the late 1960s. The first one is by far the better.

General
Friedman, Herbert *The Amazing Universe* National Geographic Society, 1975
This is an eminently readable 'special publication' from *National Geographic.*

Frontiers in Astronomy W. H. Freeman, 1970
This is a very readable book of reprints from *Scientific American,* covering most of the topics mentioned here.

Gribbin, John *Forecasts, Famines and Freezes* Wildwood, London; Walker, New York, 1976
This book includes further discussion of astronomical influences on climate.

Shklovskii, I. S. and Sagan, Carl *Intelligent Life in the Universe*
Holden-Day, New York, 1966
This book sets human life in its context within our changing universe as well as looking at the prospects for finding extraterrestrial intelligence.

The Universe Time-Life, Inc., 1974
This beautifully illustrated book has some sound information about traditional astronomy, although it is weak on the new astronomy.

Unsöld, A. *The New Cosmos* Longman, 1969
This is one of the best introductions for the serious student.

Index